Signals and Communication Technology

This series is devoted to fundamentals and applications of modern methods of signal processing and cutting-edge communication technologies. The main topics are information and signal theory, acoustical signal processing, image processing and multimedia systems, mobile and wireless communications, and computer and communication networks. Volumes in the series address researchers in academia and industrial R&D departments. The series is application-oriented. The level of presentation of each individual volume, however, depends on the subject and can range from practical to scientific.

Indexing: All books in "Signals and Communication Technology" are indexed by Scopus and zbMATH

For general information about this book series, comments or suggestions, please contact Mary James at mary.james@springer.com or Ramesh Nath Premnath at ramesh.premnath@springer.com.

Sheikh Mohammad Idrees
Roshan Jameel • Mariusz Nowostawski
Editors

Blockchain and Machine Learning Innovations

Breaking Barriers with Distributed Intelligence

Springer

Editors
Sheikh Mohammad Idrees
Department of Computer Science (IDI)
Norwegian University of Science and Technology (NTNU)
Gjøvik, Norway

Roshan Jameel
Department of Computer Science
Westford University College
Sharjah, United Arab Emirates

Mariusz Nowostawski
Department of Computer Science (IDI)
Norwegian University of Science and Technology (NTNU)
Gjøvik, Norway

ISSN 1860-4862 ISSN 1860-4870 (electronic)
Signals and Communication Technology
ISBN 978-3-031-88094-0 ISBN 978-3-031-88095-7 (eBook)
https://doi.org/10.1007/978-3-031-88095-7

This Springer imprint is published by the registered company Springer Nature Switzerland AG
The registered company address is: Gewerbestrasse 11, 6330 Cham, Switzerland

Preface

Blockchain and Machine Learning are the two revolutionary technologies that have bought changes in businesses, industries as well as our day-to-day lives. These are among the forefront technologies that open new possibilities and opportunities for diverse industries such as healthcare, education, finance, and much more. The combination of predictive capabilities of machine learning algorithms and decentralized nature of blockchain make all of this practically possible. The aim of this book *Blockchain and Machine Learning Innovations* is to provide a comprehensive resource for exploring this fascinating combination. The chapters in this book offer a diverse and insightful collection of cutting-edge research and innovative perspectives from leading experts in the field. This volume aims to showcase the synergy between blockchain and machine learning, demonstrating how these transformative technologies drive advancements across various domains.

The first chapter of this book delves into the integration of biometric authentication, blockchain, and machine learning for developing credible digital identity system. Followed by a chapter that explores the blockchain-based solutions for identifying deepfake videos, which is one of the critical issues faced by world today. Sustainable blockchain energy consumption predictions, AI-based smart farming practices, the initial chapters of the book sets foundation for understanding how these technologies can be used in addressing real-world problems. More specific use cases of blockchain and machine learning are discussed afterwards including a novel blockchain framework for wheat crop supply chain management and a chapter discussing secure architectures for e-tendering and bidding systems. Furthermore, areas like predictive analytics for smart cities and healthcare systems is also explored, covering trends in Alzheimer's disease research and designing secure frameworks for medical records. The book also addresses the areas like the role of AI and blockchain in education and their integration in the Fourth Industrial Revolution covering broader societal implications of these technological innovations.

Blockchain and Machine Learning Innovations book is designed to be an informative resource for researchers, practitioners, and enthusiasts. We extend our deepest gratitude to the contributors whose dedication and expertise have made this volume possible. Their work not only advances our understanding of blockchain and machine learning but also inspires new ideas for leveraging these technologies in meaningful ways.

Sheikh Mohammad Idrees
Roshan Jameel
Mariusz Nowostawski

Contents

Editors and Contributors

About the Editors

Sheikh Mohammad Idrees is a researcher, educator, and practitioner specializing in blockchain technologies, decentralized systems, ml, and data science. With a PhD in computer science and years of academic and industry experience, Idrees has been at the forefront of exploring innovative solutions to complex technological challenges. His expertise spans blockchain interoperability, data integrity and the integration of machine learning with decentralized technologies. As a certified Enterprise Blockchain Architect, he bridges the gap between theory and practice, mentoring students and collaborating with global researchers to advance the frontiers of technology. Currently working as a Researcher and Lecturer at NTNU, Gjøvik Norway, Idrees is committed to strengthen interdisciplinary research and innovation in emerging technologies.

Roshan Jameel is a researcher with expertise in cutting-edge technologies such as blockchain, machine learning, and cloud computing. Her academic contributions include multiple publications that explore the integration of these technologies in sectors like healthcare, energy, and education. Roshan Jameel is passionate about advancing the frontiers of knowledge through interdisciplinary research and aims to bridge theoretical advancements with practical applications to address real-world challenges.

Mariusz Nowostawski is a Professor at the Norwegian University of Science and Technology (NTNU), specializing in programming, decentralized technologies, computer science, and education. He obtained his doctorate from the University of Otago, New Zealand, where he spent 15 years as a researcher and lecturer. His MSc research focused on AI and machine learning, while his PhD work explored autonomous systems and the computational modeling of biological processes. Passionate about self-organizing systems, adaptive computation, and autonomy, Mariusz has also worked on high-performance networking applications on GPUs and multicore systems with Sun Microsystems and Oracle. Currently, his research interests include cryptocurrency privacy and anonymity, as well as blockchain forensics, in collaboration with Europol. An active and influential researcher, Mariusz continues to make valuable contributions to his fields of expertise.

Contributors

S. Aarthi SASTRA Deemed University, Thanjavur, Tamil Nadu, India

Maher Alndiwee Westford University College, Sharjah, UAE

Waseem Ahmad Bhat North Campus, University of Kashmir, Srinagar, Jammu and Kashmir, India

Varsha Bhatia ADYPU Pune, Pune, Maharashtra, India

Benita Christopher Westford University College, Sharjah, UAE

S. Dinesh SASTRA Deemed University, Thanjavur, Tamil Nadu, India

K. Geetha SASTRA Deemed University, Thanjavur, Tamil Nadu, India

Richa Gupta Amity University Uttar Pradesh, Noida, Uttar Pradesh, India

Sheikh Mohammad Idrees Department of Computer Science, Norwegian University of Science and Technology, Gjøvik, Norway

Jaweria Izhar DeMont Institute of Management and Technology, Dubai, UAE

Roshan Jameel Westford University College, Sharjah, UAE

Kanika Department of Computer Science, Chitkara University, Rajpura, Punjab, India

Department of Artificial Intelligence, Chitkara University, Rajpura, Punjab, India

Nasvi Kareem Westford University College, Sharjah, UAE

Rupinder Pal Kaur Westford University College, Sharjah, UAE

Keshav Department of Computer Science, Chitkara University, Rajpura, Punjab, India

Department of Artificial Intelligence, Chitkara University, Rajpura, Punjab, India

Sunita Kumawat Amity University Haryana, Gurgaon, Haryana, India

Halimah Abdul Manaf UUM College of Law, Government, and International Studies, Universiti Utara Malaysia, Sintok, Kedah, Malaysia

Khyati Marwah DeMont Institute of Management and Technology, Dubai, UAE

Sheetal Mavi Amity University, Noida, Uttar Pradesh, India

Deepti Mehrotra Amity University Uttar Pradesh, Noida, Uttar Pradesh, India

Neetu Mittal Amity University, Noida, Uttar Pradesh, India

Ashish Kumar Mourya Department of Information Technology, ABES Institute of Technology, Ghaziabad, Uttar Pradesh, India

Pratyusa Mukherjee School of Computer Engineering, Kalinga Institute of Industrial Technology (KIIT) Deemed to be University, Bhubaneswar, India

Sheikh Umar Mushtaq Lovely Professional University, Phagwara, Punjab, India

Dyah Mutiarin Department of Government Affairs and Administration, Universitas Muhammadiyah Yogyakarta, Yogyakarta, Indonesia

Achmad Nurmandi Department of Government Affairs and Administration, Universitas Muhammadiyah Yogyakarta, Yogyakarta, Indonesia

Aryan Rudr Pandit Department of Computer Science, Chitkara University, Rajpura, Punjab, India

Department of Artificial Intelligence, Chitkara University, Rajpura, Punjab, India

Andi Luhur Prianto Department of Government Studies, Universitas Muhammadiyah Makassar, Makassar, Indonesia

Vikas Sagar Noida Institute of Engineering and Technology, Noida, India

Pradeepta Kumar Sarangi Chitkara University School of Engineering and Technology, Chitkara University, Baddi, Himachal Pradesh, India

N. Sasikaladevi SASTRA Deemed University, Thanjavur, Tamil Nadu, India

Alisha Sikri JIMS, Rohini, Delhi, India

Arshdeep Singh Department of Computer Science, Chitkara University, Rajpura, Punjab, India

Department of Artificial Intelligence, Chitkara University, Rajpura, Punjab, India

Rashi Singh DeMont Institute of Management and Technology, Dubai, UAE

Imron Sohsan Faculty of Humanities and Social Science, Khon Kaen University, Khon Kaen, Thailand

Kanika Taneja Department of Computer Science and Engineering-Data Science, ABES Institute of Technology, Ghaziabad, Uttar Pradesh, India

Tawseef Ahmed Teli Higher Education Department, Jammu, Jammu and Kashmir, India

Bhawna Wadhwa Noida Institute of Engineering and Technology, Noida, India

Muhammad Younus Department of Government Affairs and Administration, Universitas Muhammadiyah Yogyakarta, Yogyakarta, Indonesia

Department of Product Research and Software Development, TPL Logistics Pvt Ltd, Karachi, Pakistan

CredibleIDs: Leveraging Biometric Authentication, Blockchain Technology, and Machine Learning for Enhanced Digital Identity Management Systems

Pratyusa Mukherjee

1 Introduction

In today's interconnected world, digital identity is becoming more predominant. We usually have multiple digital identities connected to our daily life, places of employment, and other activities pertaining to our careers and daily needs. A digital identity [1, 2] is a collection of credentials and identifiers expressed in an appropriate context. Identity providers have major supremacy over the existing digital identities rather than the users themselves, making it vulnerable to identity misuse. Identity owners often unknowingly reveal their credentials to register or get access for some service. However, it lacks standard guidelines specifying what data needs to be shared and stored on the Internet. This, oversharing of data also contributes to privacy breaches for the identity owner. This obviously explains why descent Digital Identity Management Systems (DIMS) should be designed to manage and secure our personal data while offering pertinent services.

Traditional password-based authentication systems [3] require users to create a secret combination of letters, numbers, and symbols to access their accounts. Enforcing password complexity requires changing passwords on a regular basis and putting lockout mechanisms after many unsuccessful login attempts. The primary advantage of these systems is their simplicity and low cost. However, they also present significant drawbacks [4], such as vulnerability to phishing attacks, brute-force

P. Mukherjee (✉)
School of Computer Engineering, Kalinga Institute of Industrial Technology (KIIT) Deemed to be University, Bhubaneswar, India
e-mail: pratyusa.mukherjeefcs@kiit.ac.in

S. M. Idrees et al. (eds.), *Blockchain and Machine Learning Innovations*, Signals and Communication Technology,
https://doi.org/10.1007/978-3-031-88095-7_1

attempts, and the risks associated with weak or reused passwords. Furthermore, users often find it difficult to remember several complicated passwords, which might lead to risky behaviors like writing them down or employing passwords that are simple to figure out. This security vs. usability trade-off emphasizes the need for stronger authentication techniques.

Biometric authentication [5] employs peculiar physical characteristics, such as fingerprints, face recognition, iris patterns, or voice-prints, to verify a user's identity. It is obviously more secure as it relies on traits that are difficult to replicate or steal compared to traditional passwords. They allow quick access to devices and applications without memorizing complex passwords. However, because biometric data cannot be altered like a password once it is compromised, there are significant privacy and data security concerns [6]. Furthermore, the technology may encounter issues with accessibility for individuals with certain disabilities and accuracy, such as false positives or negatives.

Blockchain technology [7–12] revolutionizes data management by creating a safe, decentralized, and impenetrable digital ledger. It is a distributed system [13] where duplicates of the ledger are created and maintained across a network of nodes thus rendering it highly resistant to infringing. Blockchain essentially relies on cryptographic hashing functions [14] that convert data into distinct and irrevocable string, known as a hash. Any abatement to the data results in an abruptly different hash value, thus disclosing the fact that data has been tampered with. Furthermore, digital signatures [15] are inculcated to verify the genuineness and origin of data. Blockchain networks also utilize consensus mechanisms [16] where all participants in the network consent on the new data entries added to the ledger.

By examining patterns and behaviors, machine learning [17] improves entity authentication by safely and reliably confirming identities. Anomaly detection [18] is one application in which algorithms identify typical user activity and highlight any variations that would point to illegal access. Behavioral biometrics examine mouse movements, typing habits, and other user interactions to generate a distinct profile for every person. Additionally, ML models can process and validate biometric data with increased accuracy and decreased false positives. These flexibilities thus make machine learning a strong candidate for entity authentication across a range of applications.

This chapter basically blends the nuances of biometrics, Blockchain Technology and Machine Learning to propose, CredibleIDs, an authentication scheme that leverages the strengths of these technologies to offer a promising solution to address the limitations of traditional entity authentication methods. The related work is presented systematically in section "Related Work." Section "Proposed Methodology" puts forward the proposed scheme. The security analysis is performed in section "Security Analysis of Proposed Methodology" with the conclusions drawn in section "Conclusion and Future Work."

2 Related Work

Bakre et al. [19] implemented a decentralized digital identity using blockchain to create and manage Digital Identities, authenticating log sessions and recording the transactions. Their proposed digital avatar could record and remember precise information and doings of a user from the past thus predicting what their actions would be in the future. Incorporating blockchain gave the power to own and control this identity. Similar blockchain integrated Digital Identity scheme was preferred by Buccafurri et al. [20], Mishra et al. [21], and Zwitter et al. [22]. Yuan Liu et al. [23] in their paper, proposed another blockchain-based identity management system. They discussed both the identity authentication model and reputation model in detail. Concepts of *RpCoinDay* and reputation fluctuation factor *Rpf* were also put forward by them. Their proposal made the participants maintain their personal reputation along with their user identity.

Takemiya and Vanieiev [24] presented the Sora identity system, a blockchain technology-based mobile app to create a secure protocol to store personal information in encrypted format, as well as share verifiable claims about personal information using Hyperledger Iroha. Gao et al. [25], Hamer et al. [26], and Gudala et al. [27] proposed different frameworks for identity management that also leveraged biometric authentication.

Xiaohui Yang and Wenjie Li [28] came up with a zero-knowledge-proof-based DIM scheme in blockchain which effectively hides the reveal of the ownership details. Song et al. [29] put forward a digital identity verification and management systems (DIVMS) of blockchain-based verifiable certificates (VC) that are implemented on the smart contracts of the blockchain to overcome single point failures. Bandara et al. [30] proposed a blockchain and self-sovereign identity based digital identity (KYC—Know Your Customer) platform "Casper" to address the issues on centralized identity systems.

Mamun et al. [31] implemented a blockchain-based digital identity for individuals using bio-information and ethereum smart contract. Agarkar et al. [32] presented a system that allows verifiers to authenticate documents without any disclosure of data, then share it with service providers, and access several services using their universal identity in the trusted network. Their system ensured the security of the system and privacy. In this scheme also, ethereum's smart contracts solve excessive third-party dependency, enabling users to control their identity.

Extensive review on Blockchain-Based Identity Management Systems have been provided by Liu et al. [33], Sung et al. [34], Nusantoro et al. [35], Bai et al. [36] and Huda Seyam and Adib Habbal [37] which presents a systematics analysis of the recent blockchain-based identity management solutions along with discussing their architecture, components, and features.

The applications of Machine Learning on security were discussed by Varun Shah [38]. Liu et al. [39], in their comprehensive survey article, thoroughly examined three types of interactions between privacy and machine learning: private ML,

ML-assisted privacy protection, and ML-based privacy attack, along with related security strategies.

3 Proposed Methodology

The proposed CredibleIDs scheme has been divided into several phases for the ease of explanation.

3.1 Phase 1: User Enrollment and Authentication Process

The User Enrollment process involves capturing the biometric data such as iris scan and fingerprint and extract features from both of them.

Extracting Minutiae Points from Fingerprints. After the scanner acquires the fingerprint, image processing techniques such as contrast adjustment and noise reduction are applied in order to improve the quality of the fingerprint image. The gray-scale fingerprint image is then converted into a binary image where ridges are typically represented as black color and valleys as white. There are two predominant types of minutiae. First are the terminations which are the points where a ridge ends. Second are bifurcations which means the points where a ridge splits into two. The number of crossings and its neighborhood analysis helps to identify minutiae points based on the structure of the binary image. Some of the common minutiae patterns are shown in Fig. 1. The coordinates of each minutiae point along with

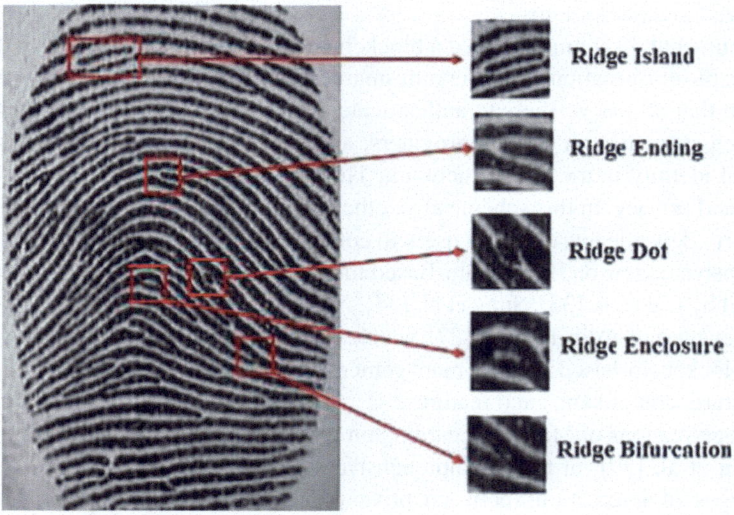

Fig. 1 Some common minutiae patterns [40, 41]

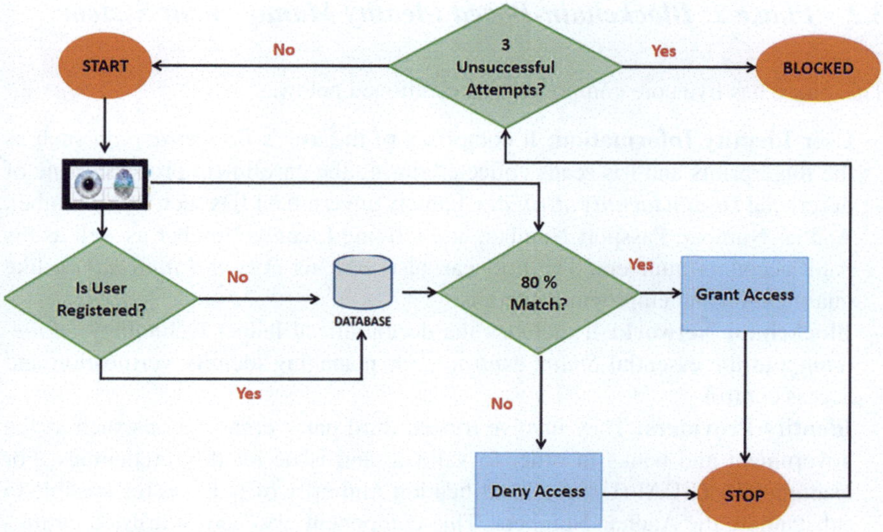

Fig. 2 Block diagram of User Enrollment and Authentication Process

additional information like angle and type are recorded in the database for later matching.

Extracting Unique Patterns from Iris Scans. An iris scanner captures a high-resolution image of the iris, ensuring proper lighting and focus. The image should be clear and free from obstructions. The captured iris image is segmented from the rest of the eye using techniques like Circular Hough Transform to identify the iris boundary. Next a polar coordinate representation of the iris is created to ensure uniformity. Finally, the extracted iris features are converted into a binary template, often referred to as an "iris code." This code is typically a compact representation of the unique patterns found in the iris. It is stored securely for subsequent comparisons during authentication.

Both the extracted features are securely stored in a database along with user ID. For User Authentication, user again presents his fingerprint and iris scan for verification. Features are extracted from the incoming fingerprint and iris scans and compared with those stored in the database. Euclidean distance formula can be used to compare the minutiae points. Pattern recognition techniques assist iris scan matching. If both matches exceed a defined of 80%, access is granted else denied. The user is notified of successful authentication else is allowed to retry for maximum three attempts. The overall block diagram of this phase is illustrated in Fig. 2.

3.2 *Phase 2: Blockchain-Based Identity Management System*

This phase has five core components as mentioned below.

- **User Identity Information:** It comprises of the user's *Biometric data* such as the fingerprints and iris scans collected during the enrollment process, some of his crucial *Digital identity attributes*, namely government IDs such PAN Number, Aadhar Number, Passport Number, and Driving License Number as well as his bank accounts numbers. The user can also add his *Personal information* like qualification and employment details.
- **Blockchain Network:** It includes the decentralized ledger technology—*Ethereum* and the essential *Smart contracts* for managing identity verification and access control.
- **Identity Providers:** They involve trusted third party organizations such as the government and banks in order to validate and issue identity credentials. For example, the UIDAI (Unique Identification Authority of India) is responsible to substantiate the Aadhar Numbers. This component also adds entrusted entities that verify the personal information like the user's graduation institute or his employee to verify his qualification and employment details before entry into the blockchain.
- **User Interface:** It incorporates all the front-end Web or mobile application for users to manage their identities as well as the Interfaces for identity providers to issue and verify credentials.
- **Storage Solutions:** It means the back-end decentralized Off-chain storage for large files pertaining to biometric data and On-chain storage for hashes of user identity data for integrity verification.

Credible ID Generation: Figure 3 illustrates the CredibleID generation procedure. Each trusted third party Identity Provider will log into the blockchain using its registered ID. Each user will have already undergone the usual process to request

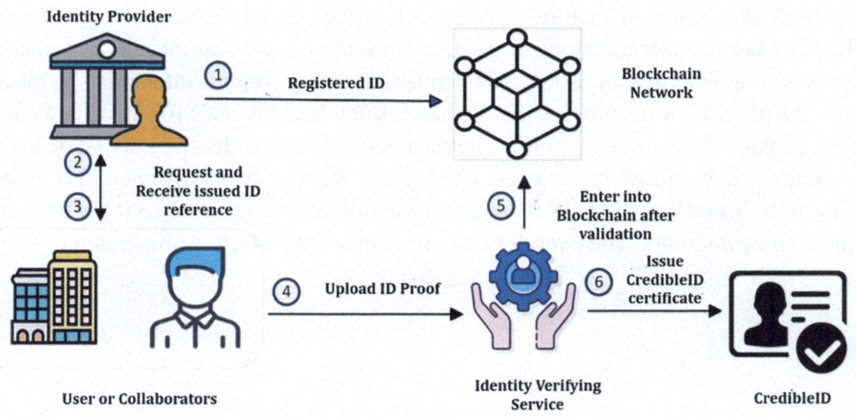

Fig. 3 Workflow model of CredibleID generation procedure

and receive their government issued ID reference number like Aadhar or Pan Number. The user will then upload this ID proof into the Identity verifying service which will validate its genuineness. If it matches, the verifying service will also upload the same into the blockchain network and return the user with his authentic all-purpose digital identity that is his CredibleID. If a collaborator (new employee or institution or application service) wants the user to verify their identity, he provides his CredibleID. The service provider again checks the credential against the blockchain. If valid, the user is granted access to services; if not, he is denied of any access.

Decentralized Identity Sharing mechanism: The proposed CredibleID also provides a Decentralized Identity Sharing mechanism where users can share specific identity attributes with third parties without revealing their entire identity. For example, they can show age without disclosing the date of birth. This is managed through Challenge Response Methodologies. Whenever a collaborator requests identity verification, the user can select which attributes to share. The user digitally signs the proof of identity attributes to ensure authenticity with his public key. The proof is then encrypted to protect it from unauthorized access. The collaborator receives the encrypted proof and verifies it against the blockchain using the public key used to sign by user and a smart contract. If verified, access to services is granted without exposing unnecessary personal data. Figure 4 illustrates the basic methodology that is followed is that the user wants to prove his identity to the verifier. The verifier asks him a challenge by encrypting it with user's public key and the user responds to it by decrypting the challenge using the corresponding private key which is known only to him. He, however, does not reveal his actual private key to the verifier. The correctness of this response shows whether the user is authentic or not. Thus authentication is guaranteed without real-time exchange of crucial information. The blockchain maintains an immutable log of all identity-related transactions. Organizations can audit access and changes to identities for compliance and security purposes.

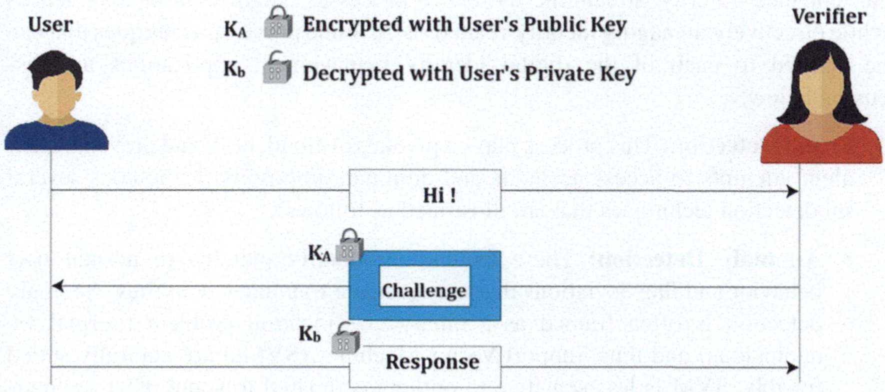

Fig. 4 Challenge response-based decentralized identity sharing mechanism

Smart Contract Deployment: Smart contracts are self-executing contracts with the terms of the agreement directly written into code. They run on a blockchain and automatically enforce and execute agreements when predetermined conditions are met. They can automate the process of issuing digital identity credentials. When a user's identity is verified by an identity provider, a smart contract can create and store a digital credential on the blockchain. Smart contracts can revolutionize access control by defining and managing access rights based on user attributes or roles. For example, a smart contract may specify that only users over a certain age can access particular services. The expiration dates and conditions for renewal of the once issued CredibleID can be facilitated through appropriate smart contracts. It also enables revoking of credentials if an identity is compromised. The list of steps is provided below.

- Step 1: The user presents his CredibleID whenever he wants to access a service.
- Step 2: The service provider calls a smart contract that checks the credential.
- Step 3: The smart contract compares the hash in the credential with the hash stored on the blockchain.
  - It verifies whether the credential is still valid (not expired or revoked).
  - It assesses whether the user has the necessary permissions for the requested service.
- Step 4: If all checks pass, the smart contract grants access and logs the transaction on the blockchain else access is denied and the reason may be logged for auditing.

3.3 Phase 3: Machine Learning Enforcement to Safeguard CredibleIDs

By leveraging machine learning techniques, digital identity management systems can enhance security, streamline the entire processes, augment user experiences while effectively managing identity-related risks. Some possible techniques that can be applied to each of the digital identity management applications are discussed below:

1. **Fraud Detection:** This process plays a pivotal role to identify and prevent fraudulent attempts to access accounts and eliminate imposters. It includes several subdetection techniques that are illustrated as follows:
   - **Anomaly Detection:** These techniques identify patterns in normal user behavior and flag deviations that may indicate fraudulent activities. Anomaly detection is often framed as a binary classification problem (normal vs. anomalous) and thus Support Vector Machines (SVMs) are naturally suited for this. SVM is less sensitive to outliers compared to some other methods and focus on maximizing the margin between the classes. This property helps

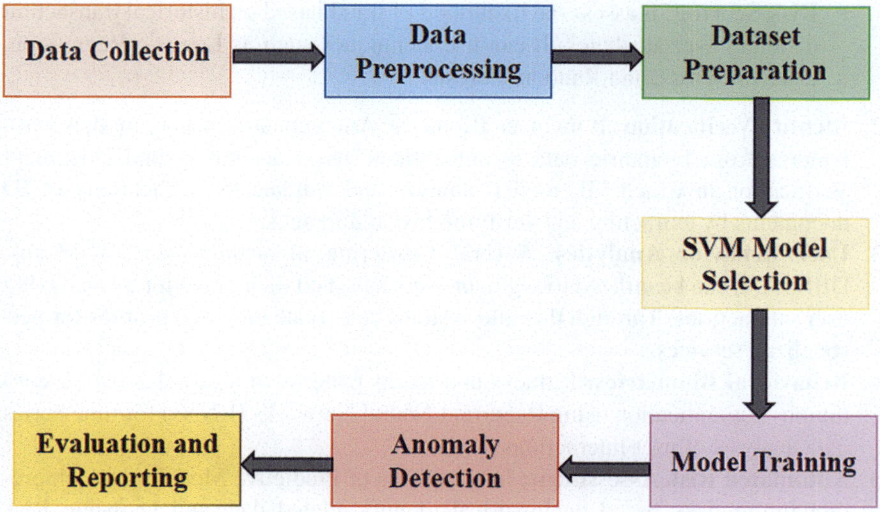

Fig. 5 Block diagram of anomaly detection process

in effectively identifying anomalies that lie far from the normal data distribution. Figure 5 presents the overall steps involved in the process.

The foremost step is to collect relevant data that may contain normal and anomalous instances. This could be from various sources like logs, sensors, or transaction records. It's essential to consider attributes that capture user behavior, access patterns, and system interactions. Some key attributes include *User Attributes* which enlists the User ID, his role or privileges and when the account was created; *Access Attributes* which takes into account the date and time of user logins, their geographic location or IP address, the type of device (e.g., desktop, mobile) and OS used during access. Time spent logged into the system is also carefully monitored, *Behavioral Attributes* which track the types of actions performed (e.g., login, logout, data access) and how often specific actions are performed (e.g., number of logins per day) and on which time of day (e.g., off-hours access), *Security Attributes* such as the number of failed login attempts before a successful login. Frequency and timestamps of password changes. This data is then preprocessed. It is then labeled with normal and anomalous instances and split training and testing sets. For SVM Model Selection, a linear kernel function can be chosen. Then the SVM model is fitted on the training dataset. The trained model then predicts anomalies on the test set. The model is finally evaluated using metrics such as precision, recall, and F1-score. It is necessary to regularly monitor the model's performance and update it as necessary with new data. Periodically retraining the model with new data to adapt to changes in patterns can ensure sustained performance. A record of the anomalies detected is kept which assists to provide insights and recommendations based on the detected anomalies.

- **Risk Scoring:** It assess the likelihood of fraud based on historical transaction data and user attributes. It can use techniques such as Logistic Regression, Decision Trees, and Random Forests.

2. **Identity Verification:** It involves Biometric Authentication where models learn features from biometric data to authenticate users accurately and Document Verification in which ML models analyze and validate the authenticity of ID documents by extracting and verifying text and images.
3. **User Behavior Analytics:** Several Clustering algorithms (e.g., K-Means, DBSCAN) can be utilized to segment users based on their behavior by analyzing user interactions. Through this, the systems can create enriched profiles for personalized services.
4. **Behavioral Biometrics:** It tracks and learns patterns in user behavior for continuous authentication using Recurrent Neural Networks (RNNs) for time-series data analysis of user interactions.
5. **Automated Risk Assessment:** In comprises of Predictive Modeling which predict future risks based on historical identity-related data and Dynamic Risk Profiling that continuously updates risk profiles based on real-time data inputs.

4 Security Analysis of Proposed Methodology

The security analysis of CredibleIDs supported by its three pillars, namely biometric authentication, Blockchain technology, and Machine Learning. The detailed analysis is provided as follows:

4.1 Analysis of Biometric Authentication

In comparison to conventional password-based verification, biometric authentication has drawn a lot of attention as a safer and more secure technique. This advantage results from a number of inherent features of biometric systems that mitigate common password-related flaws. The main elements that improve biometric authentication's safety and security are listed below.

1. **Unique Biological Traits:** Biometric authentication relies on unique physiological or behavioral characteristics of individuals, such as fingerprints, facial recognition, or voice patterns which can be shared or stolen in contrast to passwords.
2. **Inherent Uniqueness:** Each person's biometric data is distinct and cannot be easily replicated or guessed. Unlike passwords, which may be simple or shared among multiple users, biometrics are personal and tied specifically to the individual.

3. **Difficult to Duplicate:** Biometric traits cannot be forgotten or lost like passwords. The physical characteristics used in biometric authentication, such as fingerprints and facial morphology, are exceedingly difficult to forge.
4. **Phishing Resistance:** Biometric traits cannot be phished like passwords. Users cannot inadvertently provide biometric data to malicious entities, as this information is not something they enter manually.
5. **Hard to Guess:** The passwords selected by many users are often weak or predictable. In contrast, biometric data is inherently complex and unique, diminishing the likelihood of successful guessing.
6. **No Need for Memorization:** Users do not need to remember complex passwords or change them regularly, which reduces the risk of poor password management practices such as writing passwords down or reusing them across multiple accounts.
7. **Quick and Efficient:** The speed at which users can authenticate through biometrics—such as scanning a fingerprint or recognizing a face—improves the overall user experience and increases the likelihood of adopting stronger security practices.

Thus, biometric authentication systems provide a more secure alternative to traditional password-based methods due to their unique inherent traits, resistance to various forms of cyber-attacks, and improved user convenience. Although biometrics are not without vulnerabilities, their inculcation can significantly diminish the risks associated with password management, providing a more robust and efficient solution for secure access to sensitive information and systems.

4.2 Analysis of Blockchain-Based Identity Management System

Blockchain-based identity management systems offer various advantages that enhance security, privacy, and user control over digital identities. Below are the detailed advantages of such systems:

1. **Decentralization:** Data is not stored in a single central location; it is distributed over an array of computers thus diminishing the risk of data breaches.
2. **Tamper-Resistance:** Once the data is entered into a blockchain, it cannot be updated or removed without the consent of the entire network. This ensures data integrity.
3. **Cryptographic Protection:** Data is safeguarded through cryptographic schemes and challenge response mechanisms, making unauthorized access almost negligible.
4. **Self-Sovereign Identity:** Individuals have complete ownership and control over their personal data and can decide what to share and with whom to share.

5. **Minimized Data Sharing:** Users can share only necessary data without revealing sensitive personal information, escalating the privacy.
6. **Instant Verification:** Real-time verification of transactions speed up services requiring KYC (Know Your Customer) checks.
7. **Lower Verification Costs:** Automated verification through smart contracts can cut down substantial costs associated with manual verification.
8. **Reduction in Fraud Prevention Costs:** Blockchain minimizes identity theft and fraud.
9. **Universal Access:** Anyone with a mobile device can participate, enabling easier access.
10. **No Central Point of Failure:** There is no single point of failure that can be compromised, thereby discarding the chances of mass identity theft.
11. **Control Over Identity Data:** Users manage their own credentials in decentralized wallets, which curbs unauthorized third parties interferences.
12. **Regulatory Compliance:** Personal data sharing is consensual and traceable, thereby enabling organizations cater to stringent legal requirements.
13. **Data Minimization:** Blockchain's inherent design that supports data minimization.
14. **Auditable Transactions:** Every transaction can be traced back, providing a transparent audit trail that strengthens trust among users and organizations.
15. **Transparency in Operations:** Users can verify their credentials independently, fostering trust in the identity management process.
16. **Cross-Platform Usability:** One verified identity can be used across several platforms.
17. **Facilitating Smart Contracts:** Blockchain allows implementation of smart contracts that can automate various processes within identity management, thereby reducing the need for intermediaries.

Because of its benefits, blockchain-based identity management solutions hold promise as a remedy for a number of issues with conventional identity management. They are positioned as a transformative factor in the continuous evolution of digital identity management because of their capacity to improve security, privacy, efficiency, and user control.

4.3 Analysis of Machine Learning Enforcement

Compared to conventional techniques, machine learning (ML)-based intrusion and anomaly detection systems have many benefits. These benefits improve security measures' efficacy, efficiency. Below are the detailed advantages:

1. **Improved Accuracy:** Machine learning algorithms can swiftly analyze gigantic amounts of data to identify patterns and detect anomalies.
2. **Pattern Recognition:** ML algorithms can effectively recognize complex patterns in data that might be missed by traditional rule-based systems. This allows more accurate detection of intrusions and anomalies.

3. **Adaptive Learning:** As more data is ingested, ML models can improve their detection capabilities by adapting to new threats dynamically without manual updates.
4. **Instantaneous Analysis:** Real-time analysis ensures threats are identified and mitigated quickly.
5. **Automated Responses:** ML systems can integrate automated response mechanisms that can take prompt actions upon detecting threats and reduce potential damage.
6. **Handling Large Datasets:** ML algorithms are designed to work efficiently with big data, allowing organizations to analyze extensive logs and network traffic without significant performance degradation.
7. **Flexible Frameworks:** ML systems can be adapted to from small networks to large enterprises, while maintaining scalability.
8. **Contextual Awareness:** By learning from historical data, ML algorithms can better distinguish between benign and malicious behaviors, thus enhancing the precision of alerts.
9. **Continuous Improvement:** The models can be continuously trained using feedback loops to refine their detection capabilities, resulting in fewer false positives over time.
10. **Reduced Need for Manual Monitoring:** Automation through ML minimizes the need for constant human oversight, allowing security teams to focus on more complex tasks.
11. **Efficient Resource Usage:** ML systems can operate with fewer resources compared to traditional systems that require extensive manual configuration and updates.
12. **Behavioral Analysis:** Machine learning can create baselines of normal activity, enabling the detection of deviations that may indicate intrusions or anomalies.
13. **Discovery of Unknown Threats:** Unlike signature-based systems that only detect known attacks, ML can identify previously unseen threats through anomaly detection techniques.
14. **Cross-Compatibility:** ML models can complement existing security infrastructures, such as firewalls enhancing their functionality.
15. **Collaboration with AI Technologies:** ML can work alongside other AI technologies, such as natural language processing (NLP) and computer vision, to bolster security measures further.
16. **Predictive Analysis:** ML algorithms can analyze historical threats and identify potential future attacks, enabling organizations to strengthen their defenses before incidents occur.
17. **Continuous Monitoring:** These systems can continuously monitor network environments and log files for any suspicious activity, ensuring a 24/7 defensive posture.
18. **Dynamic Adaptation:** Rather than relying on static rules that need constant updating, ML systems automatically adapt to new data patterns and threats, reducing the administrative burden.

19. **Enhanced Data Monitoring:** By providing continuous and automated monitoring and reporting capabilities, ML helps organizations adhere to regulatory requirements regarding data protection and breach notifications.

The application of machine learning in intrusion and anomaly detection not only enhances detection accuracy and speed but also provides significant operational advantages, such as cost reduction, reduced false positives, and improved scalability. By leveraging these capabilities, organizations can maintain a more robust and proactive security posture in an increasingly complex threat landscape.

5 Conclusion and Future Work

CredibleIDs is a paradigm shift in security and identity verification that combines biometric authentication, blockchain-based identity management, and machine learning-based intrusion and anomaly detection. Compared to conventional techniques, biometric solutions offer a higher degree of security and user ease. These technologies efficiently prevent identity theft and expedite authentication procedures by using distinctive physical and behavioral characteristics. The application of blockchain-based identity management improves user control over personal data, security, and privacy. Blockchain promotes user agency and reduces the danger of data leaks by providing decentralized identity management. By using adaptive algorithms that learn from past data, machine learning-based intrusion detection and anomaly detection greatly enhance the detection of threats within networks. By improving detection speed and accuracy, machine learning lowers the number of false positives.

Future efforts should focus on enhancing the robustness of biometric systems against spoofing and other fraudulent activities. The integration of multi-modal biometric systems, which combine various biometric modalities, can provide increased security and user flexibility. It is necessary to improve the scalability and interoperability among different blockchain platforms. Advancements in the area of Machine Learning-Based Intrusion Detection and Anomaly Detection should prioritize the development of models that can better handle class imbalance and concept drift, ensuring effective performance in dynamic environments. Increasing the explainability of machine learning models will help bridge the gap between high detection accuracy and user trust. Furthermore, integrating natural language processing and behavioral analytics can strengthen the capabilities of intrusion detection systems. Collaborative research across these domains could lead to innovative solutions that enhance privacy, security, and user empowerment in an increasingly digital world.

References

1. Windley, P. J. (2005). *Digital identity*. O'Reilly Media, Inc..
2. Camp, J. L. (2004). Digital identity. *IEEE Technology and Society Magazine, 23*(3), 34–41.
3. Conklin, A., Dietrich, G., & Walz, D. (2004). Password-based authentication: A system perspective. *37th Annual Hawaii International Conference on System Sciences, 2004*. Proceedings of the IEEE.
4. Kaur, A. A., & Mustafa, K. K. (2019). A critical appraisal on password based authentication. *International Journal of Computer Network and Information Security, 11*(1), 47.
5. Bhattacharyya, D., et al. (2009). Biometric authentication: A review. *International Journal of u-and e-Service, Science and Technology, 2*(3), 13–28.
6. Alsaadi, I. M. (2015). Physiological biometric authentication systems, advantages, disadvantages and future development: A review. *International Journal of Scientific & Technology Research, 4*(12), 285–289.
7. Zheng, Z., et al. (2017). An overview of blockchain technology: Architecture, consensus, and future trends. *2017 IEEE international congress on big data (BigData congress)*. Ieee.
8. Mukherjee, P., & Pradhan, C. (2021). Blockchain 1.0 to blockchain 4.0—The evolutionary transformation of blockchain technology. In *Blockchain technology: Applications and challenges* (pp. 29–49). Springer International Publishing.
9. Satybaldy, A., Subedi, A., & Idrees, S. M. (2024). Decentralized key management for digital identity wallets. In *Blockchain transformations: Navigating the decentralized protocols era* (pp. 47–58). Springer Nature Switzerland.
10. Idrees, S. M., & Nowostawski, M. (Eds.). (2024). *Blockchain transformations: Navigating the decentralized protocols era*. Springer.
11. Sharma, S., Pandey, A., Sharma, V., Mishra, S., & Alkhayyat, A. (2023, November). Federated Learning and Blockchain: A Cross-Domain Convergence. *In 2023 3rd International Conference on Technological Advancements in Computational Sciences (ICTACS)* (pp. 1121–1127). IEEE.
12. Idrees, S., & Nowostawski, M. (2022). *Transformations through Blockchain technology*. Germany: Berlin/Heidelberg.
13. Kuo, T.-T., Kim, H.-E., & Ohno-Machado, L. (2017). Blockchain distributed ledger technologies for biomedical and health care applications. *Journal of the American Medical Informatics Association, 24*(6), 1211–1220.
14. Sobti, R., & Geetha, G. (2012). Cryptographic hash functions: A review. *International Journal of Computer Science Issues (IJCSI), 9*(2), 461.
15. Fang, W., et al. (2020). Digital signature scheme for information non-repudiation in blockchain: A state of the art review. *EURASIP Journal on Wireless Communications and Networking, 2020*, 1–15.
16. Sriman, B., Ganesh Kumar, S., & Shamili, P. (2021). Blockchain technology: Consensus protocol proof of work and proof of stake. In *Intelligent Computing and Applications: Proceedings of ICICA 2019*. Springer Singapore.
17. Nassif, A. B., et al. (2021). Machine learning for anomaly detection: A systematic review. *Ieee Access, 9*, 78658–78700.
18. Chandola, V., Banerjee, A., & Kumar, V. (2009). Anomaly detection: A survey. *ACM computing surveys (CSUR), 41*(3), 1–58.
19. Bakre, A., Patil, N., & Gupta, S. (2017). Implementing decentralized digital identity using blockchain. *International Journal of Engineering Technology Science and Research, 4*(10), 379–385.
20. Buccafurri, F., et al. (2018). Integrating digital identity and blockchain. On the Move to Meaningful Internet Systems. In *OTM 2018 Conferences: Confederated International Conferences: CoopIS, C&TC, and ODBASE 2018, Valletta, Malta, October 22–26, 2018, Proceedings, Part I*. Springer International Publishing.

21. Mishra, S., Chakraborty, S., Sahoo, K. S., & Bilal, M. (2023). Cogni-Sec: A secure cognitive enabled distributed reinforcement learning model for medical cyber–physical system. *Internet of Things, 24*, 100978.
22. Zwitter, A. J., Gstrein, O. J., & Yap, E. (2020). Digital identity and the blockchain: Universal identity management and the concept of the "self-sovereign" individual. *Frontiers in Blockchain, 3*, 26.
23. Liu, Y., et al. (2017). An identity management system based on blockchain. *2017 15th Annual Conference on Privacy, Security and Trust (PST)*. IEEE.
24. Takemiya, M., & Vanieiev, B. (2018). Sora identity: Secure, digital identity on the blockchain. *2018 ieee 42nd annual computer software and applications conference (compsac)*. Vol. 2. IEEE.
25. Gao, Z., et al. (2018). Blockchain-based identity management with mobile device. *Proceedings of the 1st Workshop on Cryptocurrencies and Blockchains for Distributed Systems*.
26. Hamer, T., et al. (2019) Private digital identity on blockchain.
27. Gudala, L., et al. (2022). Leveraging biometric authentication and blockchain technology for enhanced security in identity and access management systems. *Journal of Artificial Intelligence Research, 2*(2), 21–50.
28. Yang, X., & Li, W. (2020). A zero-knowledge-proof-based digital identity management scheme in blockchain. *Computers & Security, 99*, 102050.
29. Security and Communication Networks. (2024). Retracted:: Digital identity verification and management system of blockchain-based verifiable certificate with the privacy protection of identity and behavior.
30. Bandara, E., et al. (2021). A blockchain and self-sovereign identity empowered digital identity platform. 2021 International Conference on Computer Communications and Networks (ICCCN). IEEE.
31. Mamun, M. A. A., et al. (2020). A novel approach to blockchain-based digital identity system. In *Advances in Information and Communication: Proceedings of the 2020 Future of Information and Communication Conference (FICC)* (Vol. 1). Springer International Publishing.
32. Agarkar, A. A., et al. (2024). Blockchain aware decentralized identity management and access control system. *Measurement: Sensors, 31*, 101032.
33. Liu, Y., et al. (2020). Blockchain-based identity management systems: A review. *Journal of Network and Computer Applications, 166*, 102731.
34. Sung, C. S., & Park, J. Y. (2021). Understanding of blockchain-based identity management system adoption in the public sector. *Journal of Enterprise Information Management, 34*(5), 1481–1505.
35. Nusantoro, H., et al. (2021). Blockchain based authentication for identity management. *2021 9th International Conference on Cyber and IT Service Management (CITSM)*. IEEE.
36. Bai, Y., et al. (2022). Decentralized and self-sovereign identity in the era of blockchain: A survey. *2022 IEEE international conference on Blockchain (Blockchain)*. IEEE.
37. Seyam, H., & Habbal, A. (2023). A systematic review of blockchain-based identity management solutions. *International Conference on Recent Academic Studies, 1*(1), 246–253.
38. Shah, V. (2021). Machine learning algorithms for cybersecurity: Detecting and preventing threats. *Revista Espanola de Documentacion Cientifica, 15*(4), 42–66.
39. Liu, B., et al. (2021). When machine learning meets privacy: A survey and outlook. *ACM Computing Surveys (CSUR), 54*(2), 1–36.
40. Retrieved from https://www.bayometric.com/minutiae-based-extraction-fingerprint-recognition/
41. Yager, N., & Amin, A. (2004). Fingerprint verification based on minutiae features: A review. *Pattern Analysis and Applications, 7*, 94–113.

A Study of Blockchain-Based Deepfake Video Identification

Deepti Mehrotra and Richa Gupta

1 Introduction

Exploration of data on social media has raised a big concern for the privacy of the information of an individual. This concern has been magnified many folds with technology advancement, like today fake multimedia content is easily created and widely circulated on social media. A deepfake is an image or video of a modified person using deep learning technology to create realistic fake images or videos. The term deepfake is given as it is adjoined from jargon, "deep learning," and "fake." Usage of GANs has made it possible to create fake images which are so real that one cannot quickly notice any inconsistency or distortions. Deepfakes can also utilize deep learning methods such as CNN to create fake outputs. GANs are the most popular method that is used in the creation of a deepfake. Deepfakes created with these are more challenging for forensics as these can preserve facial expression, light, etc. of the original photographs [1].

Many open-source software readily available for creating deepfake images, to name a few are, face swap [2], an app used to swap faces, DeepNude [3] where individuals can superimpose the target face onto the body on another source, DeepArt [4] creates deepfake images with an objective to turn them artistic. Face swapping is a cyber threat as it can be utilized to create fake identification and affects authentication.

Unfortunately, the advancement of Artificial Intelligence (AI) and Machine Learning (ML) algorithms, are unethically used to produce misleading information about an individual. The tempered image is so convincing that people quickly trust it, and this may result in social unrest and serious issues like changing the perception about public figures and celebrities. Enhancement of web usage, especially

D. Mehrotra · R. Gupta (✉)
Amity University Uttar Pradesh, Noida, Uttar Pradesh, India

S. M. Idrees et al. (eds.), *Blockchain and Machine Learning Innovations*, Signals and Communication Technology,
https://doi.org/10.1007/978-3-031-88095-7_2

social media, unfeigned cascade propagation of false information with malicious deception, violates the individual's privacy and sovereignty. With ample media available on social media sites, deepfake seems to have an endless dataset. This content has been utilized to spread propaganda, fake news, revenge porn, etc. Hence, there is an imperative need to recognize this content to re-establish integrity and privacy so that any individual personality perception is not jeopardized. Through this application, users could create revenge porn with just a few clicks (Fig. 1).

Deepfake content has been around for a while and often catches mass attention by targeting famous people, thus making the content a Hoax, i.e., a piece of false or unverifiable information spread rapidly by Internet users [5, 6]. These messages are disseminated across social media, often intentionally and often unintentionally, aiming to trigger positive or negative emotions in the user. The users often spread this message as they believe in the information and share it with links they relate to on social media. As the issue continues to reappear and spread, several solutions have been proposed to tackle the issue. Some reasonably new solutions have applied deep learning to detect whether the image/video is fake or genuine. These models extract salient and discriminative features to figure if the given input is genuine or not [7]. Deepfake detection is classified as a binary classification problem. A problem is classified as a binary classification problem when the predefined categories are used to categorize new inputs. The predefined categories in deepfake classification are often facial features and boundaries. This often requires a massive amount

Fig. 1 A deepfake (right) of a popular Indian actor created by the FaceApp using the input image (left)

of training percentage in a large dataset to produce more accurate results. To curb the spread of hoax, steps taken determine the integrity of images or video, that is, the detection of whether they are manipulated, then identify hoax is malignant and check the author's reliability sharing the information. Many researchers have explored various deep learning algorithm to identify the image is real or forged [8–10]. Usually, these media content are circulated widely to make it viral. To predict whether the information can be a potential candidate to be categorized as viral news can be of good input for those handling the social media platform. Thus, there is need to create a platform that can track and trace the journey of media content. Many new solutions are emerging; one such solution is the employment of blockchain [11–13].

A blockchain is a system where data is stored in a secure and decentralized way [14–16]. It works like a chain of blocks, and each block holds information along with a unique code called a hash. This hash ensures the data cannot be changed. For example, a blockchain can save a video's hash at the time it is recorded to confirm its authenticity. Smart contracts can also be used to manage rules for uploading, downloading, or sharing the video, making the process reliable and efficient.

This promises more authenticity and makes it possible to improve the traceability of the video. Fast and rapid sharing of information across the internet is referred to as information getting viral. The spread of rumors, hoaxes, fake content may create awkward situation of individual and can even damage the integrity of individual. The deepfake are often created with an objective to make the content viral. The system will help to track who uploaded the video first.

This paper discusses a web-based robust platform to generate a smart contract and detecting whether the given video is a deepfake or not. With a web-based application, users will be able to use the tool as soon as they get access to a video, they want to check to provide flexible utilization. The reliability score of person uploading video is calculated and model is proposed to predict virality score of the content. The paper is further divided into five sections which include literature review followed by proposed model. In section "Implementation Sequence," implementation of the model is described followed by conclusion and future scope.

2 Literature Review

Deepfakes are AI manipulated media content made with the intention to deceive the audience by implanting a different face and voice over the original media [17]. These manipulations are much more sophisticated than other image and video manipulation software such as Photoshop and GIMP. The fact that they are much more complex in nature makes their detection even more complex. It led to approaches like Generative Adversarial Networks or GANs that exploit the inconsistencies in a deepfake video and creates deepfake detection models that provide a binary output to the passed video being "fake" or "real."

A vast number of datasets available on the internet coupled with facile reach to computation power made deepfakes much more abundant as deep learning algorithms get more accurate with large datasets and longer training duration. This is increased the deepfake content on the internet. Several methods are consistently being found to combat deepfake and fake news. Many of these newly found methods include the use of blockchain. Authors proposed a high-level structure for the prevention of fake news and used blockchain structure to tackle such news. Another way to investigate the capability of blockchain and DLTs to battle advanced duplicity, depicting the most significant applications and recognizing their fundamental open difficulties has been put forward [18]. Prashansa Agrawal et al. put forward a new concept called DeHiDe which was the name for a hybrid model to detect false news using deep learning and assigned a credibility score to the news posted [19]. Qian Chen et al. determined the integrity of fake news by including a custom Proof-of-Authority consensus algorithm, as well as a weighted-ranking algorithm, which serves as an incentive mechanism in the gamification component.

Along with deep learning models, detecting the fake video to ensure the media is real is not a difficult task if there is a credible, secure, and trusted way to trace the history of the media. While accessing the media if it is given trusted data provenance of the digital content, it is possible to track back an item in history to prove its originality and authenticity [20]. For this, consider the certificate of authenticity method which is used for physical artworks to date where they have been with past receipts checking of actual provenance can be a mundane task. Haya R. Hasan and Khaled Salah proposed a decentralized Proof of Authenticity (PoA) system using the disruptive technology blockchain [12]. Toyoda et al. argued that blockchain can provide immutable and tamperproof data and transactions in a decentralized distributed ledger [21] this ability made blockchain a crucial technology for the purpose of effective deepfake video detection.

Md. Arquam et al. [22] put forward a framework that is based on blockchain and can be employed at a peer level for the exchange of details. A block consisted of information and a chain of blocks were made. Every hub of the network engenders the data dependent on its validity to its companion nodes. The structure was assessed with a real dataset from Facebook and had an efficiency of 83%. Gipp et al. [23] focused on using blockchain only to ensure the integrity of video content, in which they used hashing on the video and then securing that hash on an immutable blockchain. Hasan et al. presented an Ethereum blockchain-based solution that establishes the authenticity of digital content by providing credible and secure traceability to a trusted artist or publishing source.

Blockchain technology combined with deep learning algorithms can be the most efficient way to tackle the increasing threat caused by deepfakes. To achieve this, one needs to understand implemented blockchain solutions available as of now. One of the propositions by Shang et al. [24] for fake news detection utilizes the idea that the source of news can be traced by keeping a record of time stamp service and the chain connections between the blocks and consequently propose a blockchain-based approach for decentralized distributed storage for tracing the origin of the news. For the same problem, Jing and Murugesan [25] presented a framework to prevent news dissemination using the blockchain. The key point to notice here is that the prime concern of the authors here is to devise an algorithm capable of creating more transparency and credibility on the internet using the blockchain concepts.

3 Methodology

The application proposed in this paper utilizes a deep learning model to detect whether an uploaded video is a forged one or genuine. The integration with blockchain technology helps in providing another layer of security by assuring that the data is unique and untampered. Proof of work conscious algorithm is used to ensure the data security is achieved. Further the blockchain technology which used distributed ledger concept ensures the tracking and tracing of the content uploaded on the application. The flow of information in the given application is given below in Fig. 2.

In the application flow, first the user will upload the video that they wish to test the authenticity of. A blockchain framework is created where corresponding to each multimedia content uploaded block is created. A unique hash for this video is generated and compared with the already existing videos. Only if the hash is unique, the video is accepted and saved by the application. This video is then passed through the deep learning model to detect if it is a deepfake or not. If the video turns out to be authentic, then it is marked as such and is displayed on the website. Otherwise, it is flagged as potentially fake, and users of the website are warned when they try to view the video. The users can also look at the profiles of other users on the platform, where the reliability of said user is displayed, that is calculated based on the history of the videos uploaded by that user.

The blockchain technology based on proof work consensus algorithm is used for checking and validating the content on the website [26]. Whenever a new video arrives, new block is created; this process is done by a proof of work protocol. The proof of work protocol is a form of hassle puzzle which requires finding hash values, which uses some predefined hash generation function.

The deep learning approach is used for identifying the fake image is ResNeXt-50 model, which is based on CNN models ResNet and VGG [27]. The face recognition is performed as preprocessing the dataset and preprocessed data is uploaded via a data loader for the training of the model. The proposed system is designed to give the optimal results; hence instead of the traditional ResNext ResNeXT-50 Architecture is used. Long short-term memory (LSTM) is a deepfake detection method that utilizes CNN (convolutional neural network) to extract temporal features in the raw video series and is shown through the sequence descriptor. The accuracy of the model is further enhanced using RMSProp optimizer [28, 29].

RMSProp optimizer is Root Mean Square Propagation, which uses a shifting mean squared gradient. For each parameter, different learning rate is used. The learning rate is modified automatically. Adagrad exhibits a radically diminishing learning rate, and this issue is resolved RMSPROP. RMSProp changes the AdaGrad in a manner how the gradient is amassed. Gradients are amassed into an

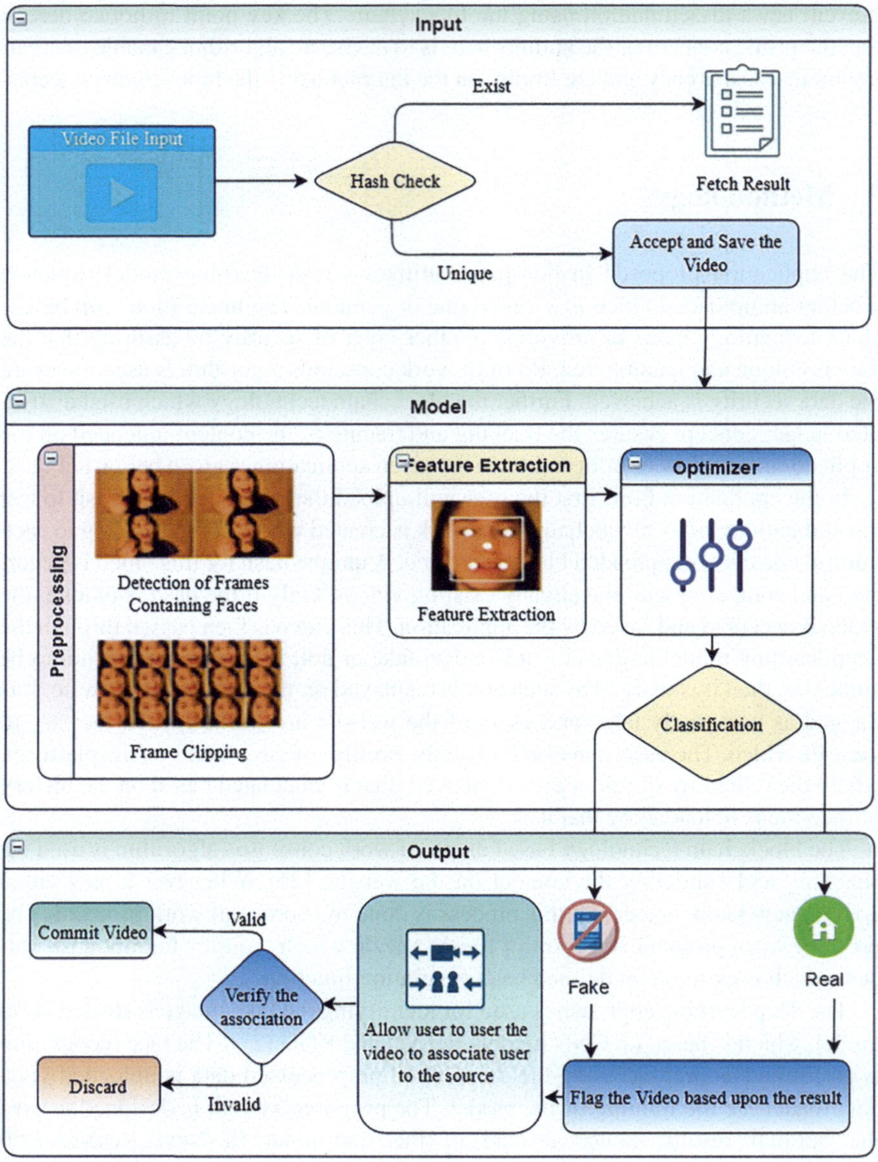

Fig. 2 Process flow diagram

exponentially weighted average. It splits the learning rate by averaging the exponential decay of squared gradients. RMSProp disposes of the past, keeps the new gradient data, and examines the RMSProp and its variations.

4 Implementation Sequence

To implement the model, the sequential flow of information from right to left is given in figure below. Figure 3 depicts that the work done in application server parameters considered include content and the person uploading image or video information. The uploaded video is forwarded to the deepfake detector to identify the genuineness of the content uploaded. The actions taken on the video depends on the type of content. The sequence of action taken is discussed in the next section and flow of information is given in Fig. 3.

The following steps are taken:

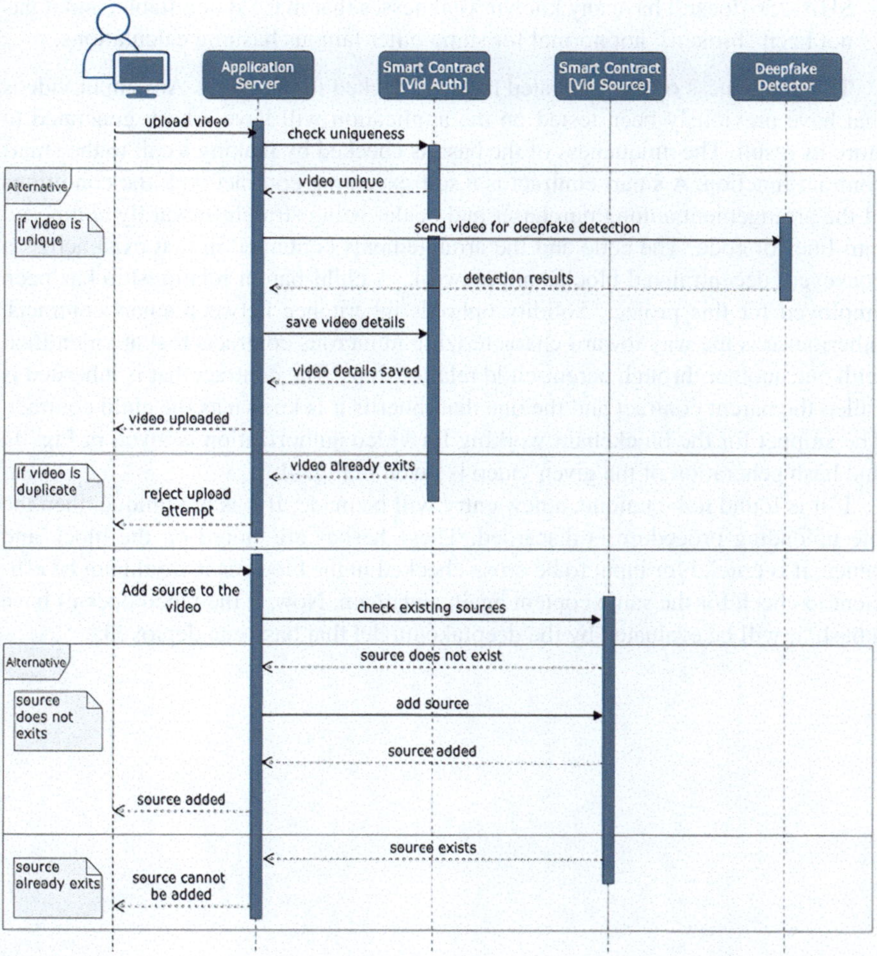

Fig. 3 Sequence diagram describing the order of interactions between the different components of the application

Step 1

First, the user uploads the input video. A hash for this video is generated. The hash function that is used for this application is SHA 256. SHA 256 is widely used in authentication and encryption protocols. Bitcoin, which is a well-known cryptocurrency utilizes SHA-256 for verifying transactions. The hashing function SHA-256 is chosen due to the following reasons:

- Collisions are unfathomably improbable: 2^{256} potential has values when utilizing SHA-256, which makes it almost inconceivable for two distinct archives to fortuitously have precisely the same hash value.
- The avalanche effect: Unlike some former hashing calculations, even a minor change to the first data totally changes the hash value; this is known as the avalanche effect.
- SHA-256 doesn't have any known weaknesses that make it unreliable, and it has not been "broken," not normal for some other famous hashing calculations.

The uniqueness of the generated hash is checked in the block. Any input videos that have previously been tested on the application will have a hash generated to store its result. The uniqueness of the hash is checked by making a call to the smart contract function. A smart contract is a self-executing contract with the conditions of the arrangement among purchaser and dealer being straightforwardly composed into lines of code. The code and the arrangements contained in that exist across a conveyed, decentralized blockchain network. A child-parent relationship has been employed for this project. Solidity upholds inheritance between smart contracts. Inheritance is the way toward characterizing numerous contracts that are identified with one another through parent-child relationships. The contract that is inherited is called the parent contract and the one that inherits it is known as the child contract. The snippet for the blockchain working for video authorization is given in Fig. 4a and hash generation of the given video is given in Fig. 4b.

If it is found to be unique, a new entry will be made. If it is not unique, then the file uploading procedure is discarded. These hashes are stored on the block and hence, it is crucial for input to be cross-checked in the block as it would not be efficient to check for the same content again and again. Now, if the video doesn't have a hash, it will be evaluated by the deepfake model that has been deployed.

```
// SPDX-License-Identifier: MIT
pragma solidity ^0.8.0

contract VidAuth
 {
   struct OwnerData
 {
      address vidOwner;
      bool exists;
   }

   mapping(string => OwnerData) ownerMap;
   mapping(address => string[]) vidsMap;

   function addNewVid(string memory vidHash)
 public
{
      require(keccak256(bytes(vidHash)) != keccak256(bytes("")));
      ownerMap[vidHash].vidOwner = msg.sender;
        ownerMap[vidHash].exists = true;
        vidsMap[msg.sender].push(vidHash);
   }

   function   isVidExists(string memory vidHash   )   view   public
returns(bool)
 {
      return ownerMap[vidHash].exists;
```

Fig. 4 Code snippet. (**a**) Video authorization solidity code snippet. (**b**) Video source solidity code snippet

```
        }

        function getOwnerFromHash(string memory vidHash ) view public
    returns(address)
    {
            return ownerMap[vidHash].vidOwner;
        }

        function getVidsFromOwner(address owner ) view public
    returns(string[] memory)
     {
            return vidsMap[owner];
        }
    }
```

4 (a): Video Authorization Solidity Code Snippet

```
// SPDX-License-Identifier: MIT
pragma solidity ^0.8.0;

contract VidSource {
   mapping(string => string[]) vidSourceMap;

   function addVidSource(string memory vidHash, string memory
vidSourceHash) public {
        require(keccak256(bytes(vidHash)) !=
keccak256(bytes("")));

        vidSourceMap[vidHash].push(vidSourceHash);
   }

   function getSourcesFromVidHash(string memory vidHash) view
public returns(string[] memory) {
       return vidSourceMap[vidHash];
   }
}
```

4 (b): Video Source Solidity Code Snippet

Fig. 4 (continued)

Step 2

When the hash generated is found to be unique, it is accepted and saved. It is then sent to the machine learning model. The model is trained to determine whether an input video is a deepfake or not. First, the input is preprocessed where it will be divided into frames and the faces that are present in these frames are cropped. There

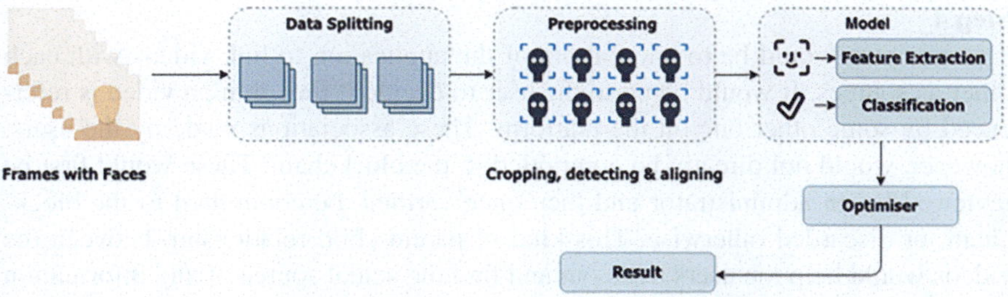

Fig. 5 Deepfake detector flow diagram

will be 300 frames present per video. This preprocessed input is then passed through the model. The sequence action taken is given in Fig. 5 below.

The model employs ResNext50 and LSTM for feature extraction and sequence processing. Both ResNext50 and LSTM are convolutional neural networks (CNN) and help with the acknowledgment of the eyes, mouth, etc. They also help in getting the bounding regression box along with the vectors. They identify facial landmarks and sites of eyes, nose, lips boundaries. LSTM is used to extract these features and evaluate the probability of an input being fake or genuine. RMSProp which is an adaptive optimizer is employed in the model. Using an optimizer will reduce the loss function. After this, the input video is labeled as either fake or genuine. Root Mean Square Propagation uses a shifting mean squared gradient. For each parameter, different learning rate is used. The learning rate is modified automatically. Adagrad exhibits a radically diminishing learning rate, and this issue is resolved RMSPROP. RMSProp changes the AdaGrad in a manner how the gradient is amassed. Gradients are amassed into an exponentially weighted average. It splits the learning rate by averaging the exponential decay of squared gradients. RMSProp disposes of the past, keeps the new gradient data, and examines the RMSProp and its variations.

Step 3

Once the input video has been classified by the AI model, the results are stored both on a database and on the blockchain. When any user of the website looks at the video, they also see how the video was classified by our model. Moreover, if someone tries posting the exact same video again, they would not be allowed to do so, and would instead be shown the results generated previously. This would preserve the uniqueness of each video on the application and also save some valuable computational resources.

Step 4
The next step would be to allow users of the application to link videos with each other as sources. It would be up to the user to decide if they think a video is referenced by some other one on the platform. These associations made by the users, however, would not directly be committed to the blockchain. These would first be reviewed by an administrator and then once verified, be committed to the blockchain, or discarded otherwise. This kind of parent-child relationship between the videos would help the users trace out and find the actual source of any information available on the platform. The visual display of the smart contract is given in Fig. 6.

Next figure is the display of authorization of video (Fig. 7).

Step 5
The video that is being checked is classified and the result is shown to the users. This result will be either fake or genuine. To gain some more insight about the authenticity of the content they are consuming, the users may look at the profile of the video's owner, where they will see their reliability score, and their past history. The users may also trace out the origin of the information in a video by successively looking at sources of the videos in the source chain of said video.

```
Smart Contract - VidSource

State:
    MAP(string, string[]) vidSourceMap

Functions:
    addVidSource
        - Input: vidHash, vidSourceHash
        - Output:
        - Begin
            vidSourceMap[vidHash].push(vidSourceHash)
          End

    getSourcesFromVidHash
        - Input: vidHash
        - Output: vidSourceMap[vidHash]
```

Fig. 6 Visual depiction of smart contract of video source

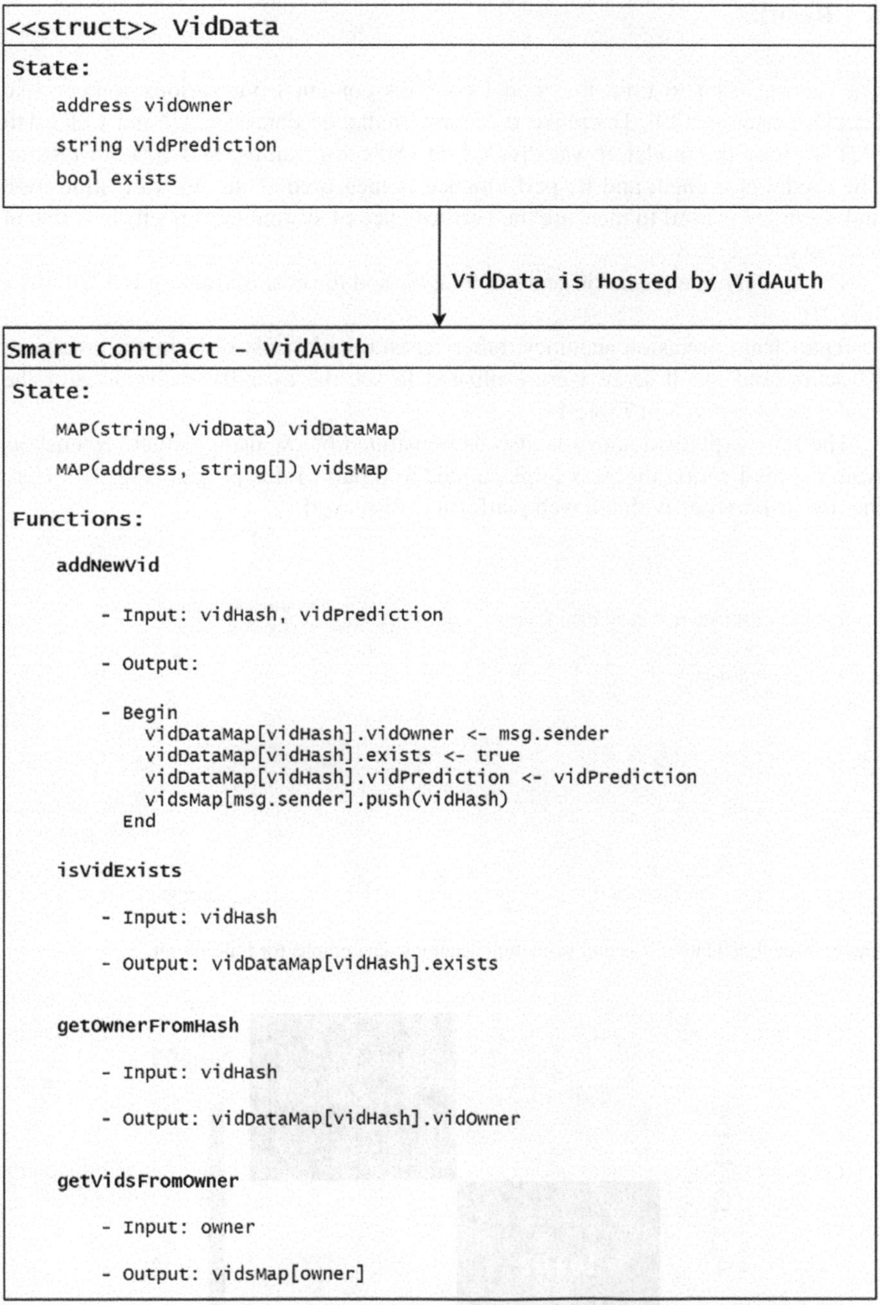

Fig. 7 Visual depiction of smart contract of video authorization

5 Result

The dataset used to train the model contains content from various sources like FaceForensics++ [30], Deepfake detection challenge dataset [31], and Celeb-DF [32]. To train the model, it was divided into 80% for training and 20% for testing. The model is trained, and its performance is measured. Training, validation loss, and accuracy is used to measure the performance of system the Epochs is varied at interval of 5 (Fig. 8).

The model has a precision score of 90.24% and the cost of training is 5220.461 s (Fig. 9).

Apart from precision accuracy, other measures like cost of training, precision, F1-score, and recall score were evaluated to see the overall performance of the model and are given in Table 1.

The flow explained above is also demonstrated below using actual screenshots from the application that was implemented as a part of this project (Fig. 10). Here the user experience with the web platform is displayed.

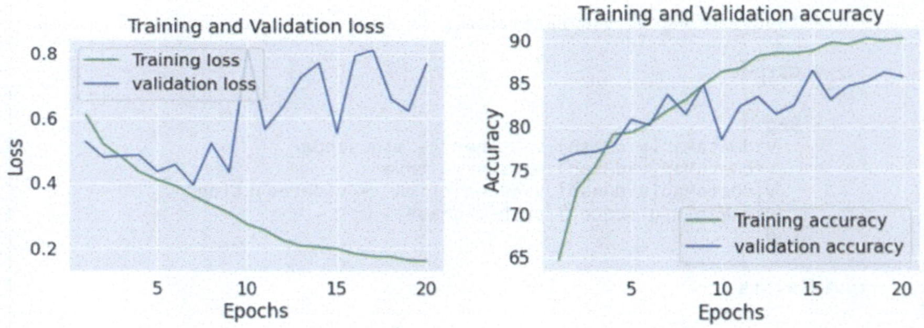

Fig. 8 Model validation loss and validation accuracy data graphs for training set

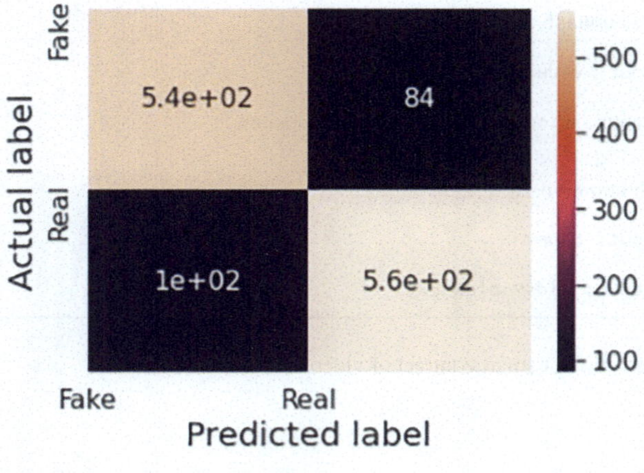

Fig. 9 Precision of the deepfake detector model

Table 1 Analysis of performance deepfake prediction model

Optimizer	Training accuracy	Testing accuracy	Cost of training	Cumulative accuracy	Precision	F1-score	Recall score
RMS Prop	90.24%	85.69%	5220.461 s	85.69%	0.835	0.865	0.897

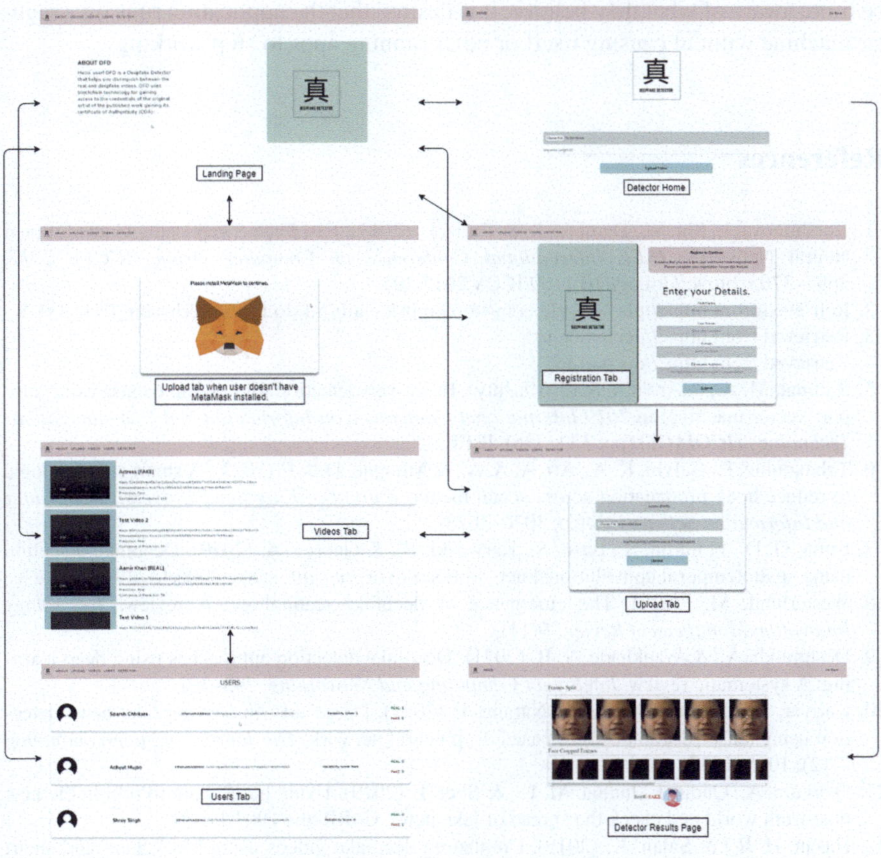

Fig. 10 User experience with blockchain-based deepfake identifier

6 Conclusion

A web application that can predict and label an input video as fake or genuine has been developed. The application uses a machine learning model with a precision score of 0.909 to successfully predict and distinguish between deepfakes and genuine videos. The application also employs blockchain to flag the result and use it directly when a user inputs the same video saving time and resources. It also preserves the uniqueness of each video that is being used as the input. The web application can successfully be used to verify the genuineness of a video and help users know how reliable the video is. Once users register their Ethereum accounts with the platform, one can upload videos themselves and further cite already existing videos as sources.

The crucial task of deepfake detection is still under development. Ways to analyze a deepfake video, as well as the deepfake audio, are constantly being searched and tested. Since the whole point of detecting deepfakes is to help the general public be more aware of what they believe. It is desired that the application runs on a regular machine without causing itself or other running apps to stop working.

References

1. Korshunova, I., Shi, W., Dambre, J., & Theis, L. (2017). Fast Face-Swap Using Convolutional Neural Networks. *IEEE International Conference on Computer Vision (ICCV), 2017*, 3697–3705. https://doi.org/10.1109/ICCV.2017.397
2. Retrieved from https://play.google.com/store/apps/details?id=io.faceapp&hl=en_IN&gl=US
3. Retrieved from https://deepnude.to/
4. Retrieved from https://deepart.io
5. Rahmat, M. A., & Areni, I. S. (2019, July). Hoax web detection for news in bahasa using support vector machine. In *2019 International Conference on Information and Communications Technology (ICOIACT)* (pp. 332–336). IEEE.
6. Tchakounté, F., Calvin, K. A., Ari, A. A. A., & Mbogne, D. J. F. (2022). A smart contract logic to reduce hoax propagation across social media. *Journal of King Saud University-Computer and Information Sciences, 34*(6), 3070–3078.
7. Lima, O. D., Franklin, S., Basu, S., Karwoski, B., & George, A. (2020). Deepfake detection using spatiotemporal convolutional net- works. *arXiv preprint*, arXiv:2006.14749.
8. Westerlund, M. (2019). The emergence of deepfake technology: A review. *Technology Innovation Management Review, 9*(11).
9. Deshmukh, A., & Wankhade, S. B. (2021). Deepfake detection approaches using deep learning: A systematic review. *Intelligent Computing and Networking*, 293–302.
10. Kaliyar, R. K., Goswami, A., & Narang, P. (2021). DeepFakE: improving fake news detection using tensor decomposition-based deep neural network. *The Journal of Supercomputing, 77*(2), 1015–1037.
11. Qayyum, A., Qadir, J., Janjua, M. U., & Sher, F. (2019). Using blockchain to rein in the new post-truth world and check the spread of fake news. CoRR abs/1903.11899.
12. Hasan, H. R., & Salah, K. (2019). Combating deepfake videos using blockchain and smart contracts. *IEEE Access, 7*, 41596–41606. https://doi.org/10.1109/ACCESS.2019.2905689
13. Yazdinejad, A., Parizi, R. M., Srivastava, G., & Dehghantanha, A. (2020, December). Making sense of blockchain for AI deepfakes technology. In *2020 IEEE Globecom Workshops (GC Wkshps)* (pp. 1–6). IEEE.
14. Idrees, S. M., & Nowostawski, M. (Eds.). (2024). *Blockchain transformations: Navigating the decentralized protocols era*. Springer.
15. Satybaldy, A., Subedi, A., & Idrees, S. M. (2024). Decentralized key management for digital identity wallets. In *Blockchain transformations: Navigating the decentralized protocols era* (pp. 47–58). Cham.
16. Singh, I., Kaur, A., Agarwal, P., & Idrees, S. M. (2024). Enhancing security and transparency in online voting through blockchain decentralization. *SN Computer Science, 5*(7), 921.
17. Balasubramaniam, P., & Sagadevan, G. (2021). Converging blockchain and AI technology-based automated and decentralized (A&D) trust management system using face detection. *International Research Journal on Advanced Science Hub, 3*, 11–15.

18. DeHiDe: Deep learning-based hybrid model to detect fake news using blockchain*.
19. Agrawal, P, Anjana, P. S., & Peri, S. An incentive-aware blockchain-based solution for internet of fake media things.
20. Floridi, L. (2018). Artificial intelligence, deepfakes and a future of ectypes. *Philosophy Technol., 31*(3), 317–321.
21. Toyoda, K., Mathiopoulos, P. T., Sasase, I., & Ohtsuki, T. (2017). A novel blockchain-based product ownership management system (POMS) for anti-counterfeits in the post supply chain. *IEEE Access, 5*, 17465–17477.
22. Arquam, M., Singh, A., & Sharma, R. (2018). A blockchain based secure and trusted framework for information propagation on online social networks. *CoRR*, abs/1812.10508.
23. Gipp, B., Kosti, J., & Breitinger, C. (2016). Securing video integrity using decentralized trusted timestamping on the bitcoin blockchain. *Proceeding of the MCIS*, p. 51.
24. Shang, W., Liu, M., Lin, W., & Jia, M. (2018). Tracing the source of news based on blockchain. In: *Proceeding of the IEEE/ACIS 17th IEEE/ACIS International Conference on Computer and Information Science*, pp. 377–381.
25. Jing, T. W., & Murugesan, R. K. (2018). A theoretical framework to build trust and prevent fake news in social media using blockchain. In *Proceeding of the International Conference of Reliable Information and Communication Technology*, (pp. 955–962).
26. Mercan, S., Cebe, M., Aygun, R. S., Akkaya, K., Toussaint, E., & Danko, D. (2021). Blockchain-based video forensics and integrity verification framework for wireless Internet-of-Things devices. *Security and Privacy, 4*(2), e143.
27. Xie, S., Girshick, R., Dollár, P., Tu, Z., & He, K. (2017). Aggregated residual transformations for deep neural networks. *Proceedings of the IEEE conference on computer vision and pattern recognition* (pp. 1492–1500).
28. Zhang, W., Zhao, C., & Li, Y. (2020). A novel counterfeit feature extraction technique for exposing face-swap images based on deep learning and error level analysis. *Entropy, 22*(2), 249.
29. Kubanek, M., Bartłomiejczyk, K., & Bobulski, J. (2019, May). Detection of artificial images and changes in real images using convolutional neural networks. In *Computational intelligence in security for information systems conference* (pp. 197–207). Springer.
30. Retrieved from https://github.com/ondyaari/FaceForensics
31. Retrieved from https://www.kaggle.com/c/deepfake-detection-challenge/data
32. Retrieved from http://www.cs.albany.edu/~lsw/celeb-deepfakeforensics.html

Predicting Blockchain Energy Consumption: A Step Towards a Sustainable Future

Aryan Rudr Pandit, Arshdeep Singh, Keshav, and Kanika

1 Introduction

The rapid growth of popularity in the field of blockchain, particularly in the realm of cryptocurrency mining, has triggered significant concerns regarding the energy consumption as it takes a whole lot of energy and can have a carbon footprint greater than that of many countries [1]. The mechanism referred to as Proof-of-Work (PoW) is still the most commonly used algorithm in numerous block chains even though it is currently known to be highly consumptive of resources. Large-scale mining in the world has seen mining operations use energy as is being used in several countries. For instance, Bitcoin mining itself is expected to take more than 100 TWh per year by 2024, which will be used to power more than the consumption of several countries [2].

Such energy consumption pattern entails disastrous environmental consequences. The continuous operation of blockchain nodes, driven largely by fossil fuel-based energy like coal-based electrical energy, deepens carbon emissions and accelerates climate change [3]. In parallel, the current energy networks around the world are experiencing revolutionary changes with changes in the share of RES (renewable energy sources) including wind and solar energy. However, integrating blockchain's energy-intensive operations into these more sustainable systems presents new challenges that are important to be addressed [4] (Fig. 1).

In this chapter, we analyze the energy consumption patterns of Bitcoin and other cryptocurrencies which are viable and have effect on the energy consumption

A. R. Pandit · A. Singh · Keshav · Kanika (✉)
Chitkara University Institute of Engineering & Technology, Department of Computer Science & Engineering, Punjab, India
e-mail: aryan3794.beai23@chitkara.edu.in; arshdeep3790.beai23@chitkara.edu.in; keshav3858.beai23@chitkara.edu.in; kanika.1272@chitkara.edu.in

S. M. Idrees et al. (eds.), *Blockchain and Machine Learning Innovations*, Signals and Communication Technology,
https://doi.org/10.1007/978-3-031-88095-7_3

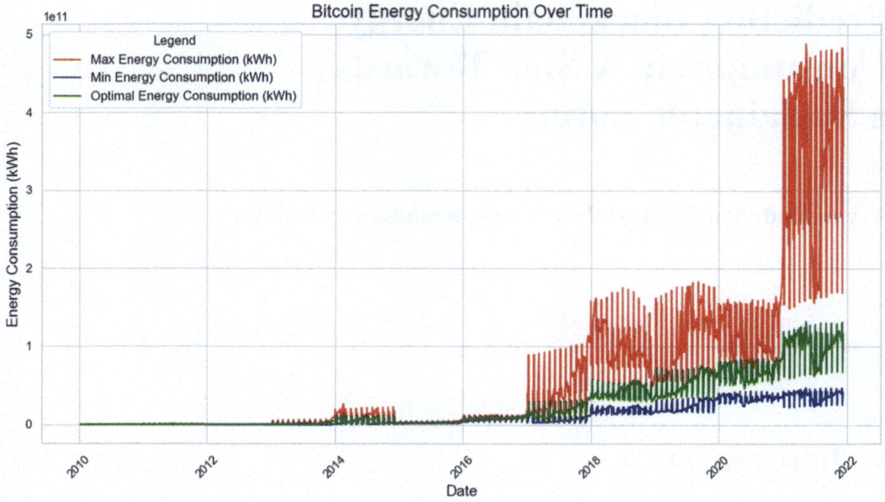

Fig. 1 Trend of Bitcoin mining energy consumption over time (2010–2022): An analysis of maximum, minimum, and optimal energy from combined energy sources

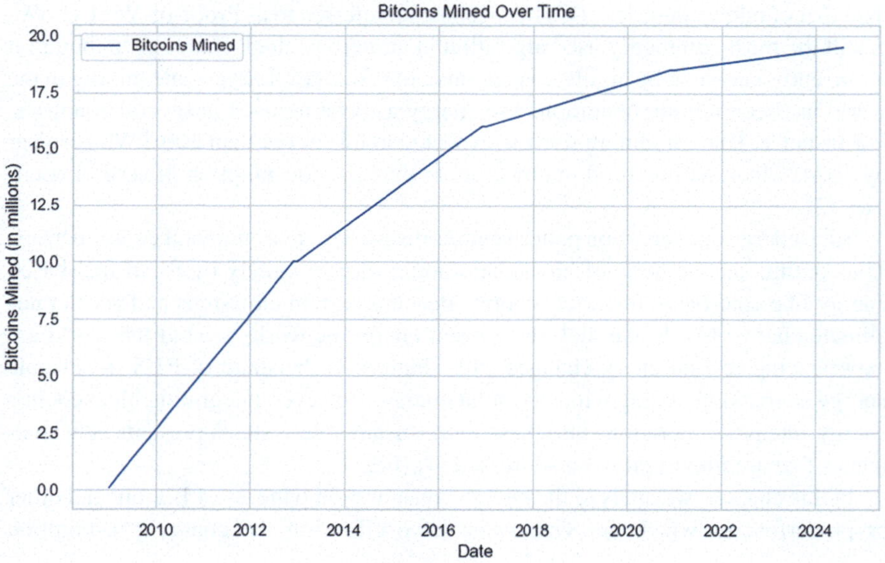

Fig. 2 Number of Bitcoins mined (in millions) since 2010

demography right from their inception in 2009 through 2024 that is shown in Fig. 2 to identify and predict the future of energy consumption patterns of blockchain mining.

Consequently, by applying machine learning methods, we predict the future values of energy consumption with regard to mining difficulty, evolution of mining hardware, development of the current network system, and associated market conditions [5]. Our analysis will not only project future energy demands but also explore the results of different strategies for mitigating this consumption through more energy efficient algorithms, such as Proof-of-Stake (PoS), Delegated Proof-of-Stake (DPoS), and Practical Byzantine Fault Tolerance (PBFT), along with the adoption of clean energy sources in mining operations other than fossil fuel-based electric energy [6].

By highlighting potential solutions in a form of different mining methods, this chapter aims to contribute to the ongoing discourse on achieving a more sustainable blockchain ecosystem to help our tomorrow in line with global energy transition goals and sustainable development efforts which can be stable even for the future. In addition, they endeavor to understand decentralized digital systems of blockchain on energy management such as smart grid technology and peer-to-peer trading that has the potential to revolutionize energy systems by increasing reliability, effectiveness and reduction in cost [7]. Hence, it will be a great foot forward to the suitable future. This problem of high energy consumption due to crypto mining shall not do any harm to our beautiful tomorrow [8].

The chapter is organized in five more sections. Section "Previous Work" explores the work done so far in the field of blockchain energy consumption. In section "Analysis and Prediction of Energy Consumption," we analyze the energy consumption over the years to depict the Bitcoin mining trend in the past decade. Here, we also expound the process and machine learning models used for predicting the energy requirements. In the next section, we elaborate the result of predictions using the five machine learning models. In section "Discussion," we discuss the implications of the analysis and outcomes of the prediction. In the end, we conclude with some future possibilities.

2 Previous Work

Through exploring various dimensions in this review, the authors seek to look at the positive and negative sides of blockchain in the energy management and renewable systems domain. In this section, we cover and explore the concurrent studies that deal with the applications, challenges, and future directions of blockchain in the energy field.

While blockchain proposes abundant benefits, several studies discuss its challenges in the energy perspective. Wang et al. [9] analyze loopholes in present-generation energy systems, examining technical obstacles like energy efficiency and scalability. They suggest new ideas like shifting the ledger transaction to Directed Acyclic Graphs (DAGs). Unlike legacy blockchain, in DAG-based ledgers transactions are organized as a graph instead of sequential blocks. It will translate the transaction as a reference to previous blocks and eliminate blocks and miners as a way to increase scalability and improve its efficiency and apply post-quantum to boost the character of security. Post-quantum cryptography enhances the security of blockchain by filling the gap that quantum computers leave in Rivest-Shamir-Adleman (RSA) and Elliptic Curve Cryptography (ECC)-based cryptography. Quantum-resistant algorithms include lattice-based, hash-based, and code-based algorithms which do not give any advantage to quantum attacks and help for safe integrity and confidentiality of blockchain transactions and data. These algorithms make blockchain systems future proof for creating secure smart contracts, transaction validations, and protecting long-term sensitive data.

While these advancements can help link blockchain to the rising energy sector's needs, regulatory and technical bottlenecks stay. This is because blockchain is not easily compatible with the actual energy system. Following a similar line of thought, Sahebi et al. [10] also focus on barriers blocking the mass adoption of blockchain in the renewable energy supply chain. This paper addresses key barriers to blockchain adoption in renewable energy supply chains (RESC), defined as regulatory, institutional, and technological challenges. Compliance and governance are also made difficult by the lack of adapted frameworks for decentralized systems; the regulatory hurdles come from that very point. Limited industrial development decentralized knowledge sharing, and lack of managerial expertise create barriers for institutions to integrate the blockchain. This means a policy change must happen to encourage the use of this technology in renewable energy administration. The analysis finds that a holistic solution must also include technical solutions, as examples in the photos showcase, and regulatory solutions.

Despite these problems, few researchers have tried different approaches to investigate solutions for combat through better management of energy consumption by blockchain technology. An example of this is a combination of AI and blockchain systems. In 2020, Kumari et al. [11] in their study say that Blockchain can be combined with AI for efficient energy management. By using AI-driven classification techniques, the model proposed in the study improves energy demand prediction and facilitates secure energy trading in smart grids. The integration of AI with blockchain is proposed as a pathway to create reliable, efficient, and autonomous energy ecosystems. Luo and Mahdjoubi [12] extend this integration concept by proposing a framework combining blockchain with machine learning to manage energy in domestic buildings. Energy use is optimized and CO_2 emissions are reduced via their model, which consists of LSTM for energy consumption prediction and Particle Swarm Optimization (PSO) model for scheduling. While we are unable to determine whether the combination of AI and blockchain would deliver better results compared to other future forecasting methods for energy management, Jin et al. [13] has illustrated the potential of such an application combining AI and blockchain for managing the complex multi-energy demands in a decentralized manner achieving cost reduction and sustainability. With the combination of federated learning and fuzzy Q learning with blockchain, they provide an optimized model for energy use in hybrid vehicles to contribute to energy recovery and reduce fuel consumption. In this paper, we illustrate how blockchain can be adapted to transport applications, especially to the distributed energy resources such as electric vehicles [Jin et al. [13]].

Another important aspect of energy management is for decentralized energy communities. Nykyri et al. [14] propose a blockchain-based solution of balancing energy consumption in residential areas. According to their study, shared photovoltaic (PV) systems and blockchain allow for self-sufficiency and efficient energy management. Self-consumption rates are improved and energy surpluses are redistributed among community residents not back to the utility grid. It is a means to show how blockchain can enable local energy trading in order to improve and optimize the energy sharing in multi-residential buildings. Gill et al. [15] believes blockchain can provide cloud computing service which supports integration with IoT and AI within decentralized applications. Secondly, their review of 140 studies highlight how blockchain, IoT, and AI can solve the security and efficiency issues for cloud-based energy applications. Proposing the framework, the authors emphasize blockchain's capacity to ensure data flows security and ensure energy resilience of cloud computing, key to decentralized systems, and Andoni et al. [16] consider wider impact of blockchain in the energy market. Their review of blockchain projects highlights technology's promise to promote transparency, secure transactions, energy democratization especially for the support of small renewable generators. This sheds light on the regulatory and scalability hurdles for block chain to transition to mass adoption in the energy space [Andoni et al. [16]].

In recent works, Ma et al. [17] addresses how blockchain technology is playing an important role in the energy sector, analyzing historical trends in energy consumption. In the work, they demonstrate the patterns in the blockchain, which can broadly point to the direction and dominant trends. In addition, the study reveals many trends in the integration of blockchain in the energy sector. Approaches to research have shifted from engineering to a more management oriented bent, with an increasing focus on governance and market structure. Among them are blockchain's support of decentralized energy transaction systems, renewable energy adoption, and smart grid functionality. But there continue to be challenges, business risks, regulatory delays, and high energy consumption. Fragmented research themes are highlighted in order to initiate more macro-level/holistic studies to address these issues in order to enable future research in the energy blockchain space. Zhao et al. [18] on the other side explores blockchain's role in localized energy trading, especially within smart communities. This study unveils a blockchain-enabled framework for peer-to-peer energy exchanges in prominent residential areas. Results determine that these decentralized models enhance self-sufficiency, allow surplus energy to be sold, and reduce dependency on centralized grid upgrades. This research and findings highlight blockchain's capability to create sustainable, autonomous energy communities, supporting clean energy distribution and promoting energy democratization.

While these studies help illuminate blockchain applications to energy, some parts remain underexplored. For instance, Zhao et al. [18] advocate on further investigation of economic feasibility for large-scale decentralized energy markets and lifecycle costs of renewable infrastructure alike. Likewise, Kumari et al. [11] suggest enhanced cybersecurity protocols to deal with privacy-related issues in the AI-integrated blockchain systems. According to Wang et al. [9], regulatory frameworks that encompass blockchain's decentralization are required in global energy markets, and Luo and Mahdjoubi [12] recommend testing blockchain and AI integration in different energy settings to examine the dynamics of the interplay between these technologies. Secondly, future research can also consider computational costs and infrastructure scalability to keep blockchain compatible with the quickly changing demands of the energy sector.

3 Analysis and Prediction of Energy Consumption

The primary objective of this research is to predict Bitcoin mining energy consumption accurately. By utilizing a range of machine learning models, this study aims to minimize energy usage forecasts and identify future consumption patterns that can guide sustainable practices in blockchain mining.

The study used two datasets on energy consumption related to Bitcoin mining. The first dataset (4159 × 16) [19] recorded daily energy metrics, including estimated energy usage and various energy footprints. To improve the accuracy and efficiency, a second dataset (138 × 31) [19] was created by aggregating this data monthly. The monthly data provides smoothed and more manageable insights for model training and testing.

From the plethora of machine learning models, we choose three most efficient models to predict the upcoming range of energy consumption due to blockchain mining-ARIMA (Autoregressive Integrated Moving Average) whose result can be seen in Table 1, XGBoost (Extreme Gradient Boosting) in Table 2 and Random Forest in Table 3. ARIMA is a statistical model known for its unique capability in time series forecasting, especially when handling stationary data [20]. Similarly, XGBoost is also good at handling complex relationships in data and for its applicability in structured, temporal datasets [[21]]. The third model we apply is Random Forest. Random Forest helps avoid overfitting and performs well on high-dimensional datasets [22]. The models are evaluated using the following three error

Table 1 ARIMA results table

Dataset	Dataset 1	Dataset 2
Training data split	70%	80%
Testing data split	30%	20%
ARIMA parameters (p, d, q)	(5, 0, 5)	(2,2,3)
Mean absolute error (MAE)	33,225,120,048.92	8,613,527,409.47
Root mean squared error (RMSE)	42,120,050,718.79	10,641,729,835.38
Mean absolute percentage error (MAPE)	40.99%	12.81%
Forecasting implication	Significant error observed	Reduced errors
Potential improvements	Fine-tuning or alternative models	Enhanced accuracy and data optimization

Table 2 XGBoost results table

Dataset	Dataset 1	Dataset 2
Mean absolute error (MAE)	19.99 * 10^9 (19.99 billion)	19.09 * 10^9 (19.09 billion)
Root mean squared error (RMSE)	28.25 billion	26.21 billion
Mean absolute percentage error (MAPE)	20.21%	19.69%
R^2 (coefficient of determination)	0.85	0.88
Potential improvements	Hyperparameter tuning and feature engineering	Hyperparameter tuning, cross-validation, and ensemble methods

Table 3 Random Forest result table

Dataset	Dataset 1	Dataset 2
Mean absolute error (MAE)	34,784,137.45	796,700,000
Root mean squared error (RMSE)	100.145 million	1.35 billion
Mean absolute percentage error (MAPE)	3.20%	7.99%
Interpretation	MAE indicates smaller prediction errors with a better fit to actual values	Average deviation of ~796.7 million. A relatively low error margin
Efficiency	Shows better accuracy with smaller deviations and more stable predictions	Shows reasonable accuracy with occasional large discrepancies
Potential improvements	Fine-tuning could lead to marginal improvements, but the model is already quite effective	Further tuning or hyperparameter adjustments could improve accuracy

metrics: Mean Absolute Error (MAE), which measures the average error magnitude, indicating general accuracy. The formula is given by Eq. (1).

$$\text{MAE} = \frac{1}{n}\sum_{\text{i}=1}^{n}\#\text{Actual}(i) - \text{Forcasted}(i)\# \tag{1}$$

where "n" is the total number of data points in the dataset, Actual(i) represents the actual observed value at the ith data point, "Forecast(i)" is the predicted or forecasted value at the ith data point, Σ (Summation) represents the sum of the absolute differences between actual and predicted values across all data points from 1 to n [23, 24].

Similarly, Root Mean Squared Error (RMSE) highlights larger errors, giving insight into significant prediction deviations. RMSE is calculated using Eq. (2).

$$\text{RMSE} = \sqrt{\frac{1}{n}\sum_{i=1}^{n}\left(\text{Actual}(i) - \text{Forcasted}(i)\right)^2} \tag{2}$$

where "n" is the total number of data points, "Actual(i)" is the observed value, "Forecast(i)" is the forecasted value, and Σ represents the sum of the squared differences between actual and predicted values across all data points [25].

Another metrics is Mean Absolute Percentage Error (MAPE); it normalizes errors as a percentage, allowing relative accuracy measurement, and is given by Eq. (3).

$$\text{MAPE} = \frac{1}{n}\sum_{i=1}^{n}\frac{\#\text{Actual}(i) - \text{Forcasted}(i)\# \times 100}{\text{Actual}(i)} \tag{3}$$

where "n" represents the number of data points, "Actual(i)" the observed value, "Forecast(i)" the forecasted value, Σ the summation across all data points, || indicates absolute error, and ×100 converts the error to a percentage [26].

3.1 Model Selection

3.1.1 ARIMA Model

ARIMA (Autoregressive Integrated Moving Average) refers to an autoregressive, integrated, and moving average which is one of the most common methods of forecasting the general value of observations in time. It is defined by three key parameters [27]. These are (p), (d), and (q) estimated by using the grid searching method as well as by the ADF stationarity test. The parameter p is the number of lagged observations or autoregressive terms in the model and the notation d is the amount of differencing applied to the dataset in order to make it stationary. The value of q, in the context of moving average, indicates the number of past forecast errors incorporated in the model. In practice, p and q are often integers within the region of 0–5 as increasing these values beyond this precludes a lift in model performance due to overfitting as a result of increased model complexity. The value of the differencing parameter d is often chosen in interval [0, 2], more frequently differentiated data can smooth out some useful features of time series. Thus, by choosing the appropriate values of the given parameters within these intervals, there is possibility to analyze the character of the time series without the introduction of the large additional complexity and without the loss of important trends [28, 29].

3.1.2 XGBoost Model

XGBoost is a gradient boosting in which the models are built for accuracy on non-linear, structured data related or connected datasets. Parameters: Hyperparameters include number of trees, number of iterations, learning rate, max depth of the tree = (number_of_trees = 1000), (rate_of_learning = 0.01), (maximum depth of the tree = 5). In XGBoost, these parameters play important roles in controlling the training process and performance of the model: The basis of the ensemble model is the number of trees and corresponds to the estimators, *n*_estimators = 1000. The larger value implies here that the model has the ability to capture a feature of a higher order and however, the time taken in training is more and always prone to overfitting parameters: (number_of_trees = 1000), (rate_of_learning = 0.01), (maximum depth of the tree = 5). In XGBoost, these parameters play important roles in controlling the training process and performance of the model: The number of trees that form the ensemble model is the estimators and is defined by *n*_estimators = 1000. The larger value means that the model has the ability to learn higher order features; however, the time taken for training is more and may lead to overfitting. By setting the parameter *n*_estimators like 1000, the model can create an instance of up to one thousand trees which will utilize mistakes of previous trees in improving performance. In the XGBoost, the default value of the learning_rate = 0.01 which bears the shorthand "eta," refers to the weights allocated to individual trees in a model. They fixed a standard value of 0.01 for each so that each step is small and the learning rate is therefore low but is able to generalize. But, when learning rate is low, the number of trees to give a good result are many, default is 100, *n*_estimators = *X*. The maximum tree depth a tree is set to have is at max_depth = 5; this usually ranges from 5 to 30. Deeper model enables modeling higher level interactions but at the same time incurs overfitting. With the help of an example when max_depth = 5, the main idea is explained that it helps to set limitations for the depth of the tree and, therefore, provides a sufficient amount of depth in between considering every incoming feature separately and stressing on this particular feature only excessively.

3.1.3 Random Forest Model

Random Forest enhances the applicability of decision trees in that it allows an ensemble of decision trees to make a prediction in order to increase accuracy but reduce bias. Parameters: It is also possible to adjust the settings to the most important parameters: (*n*_estimators = 100) and (max_depth = 10). In a Random Forest model, these parameters play an important role in shaping the complexity and robustness of the ensemble: *n*_estimators = 100 stands for the number of decision trees needed to establish on the mentioned forest. By setting *n*_estimators to 100, what it means is that the proposed model is 100 trees which are developed from a randomly chosen data sample as well as features. If more trees are created, accuracy increases and there's less of an opportunity for a single tree to over-fit on the data, but computation time will be higher. The Random Forest with 100 trees as trained

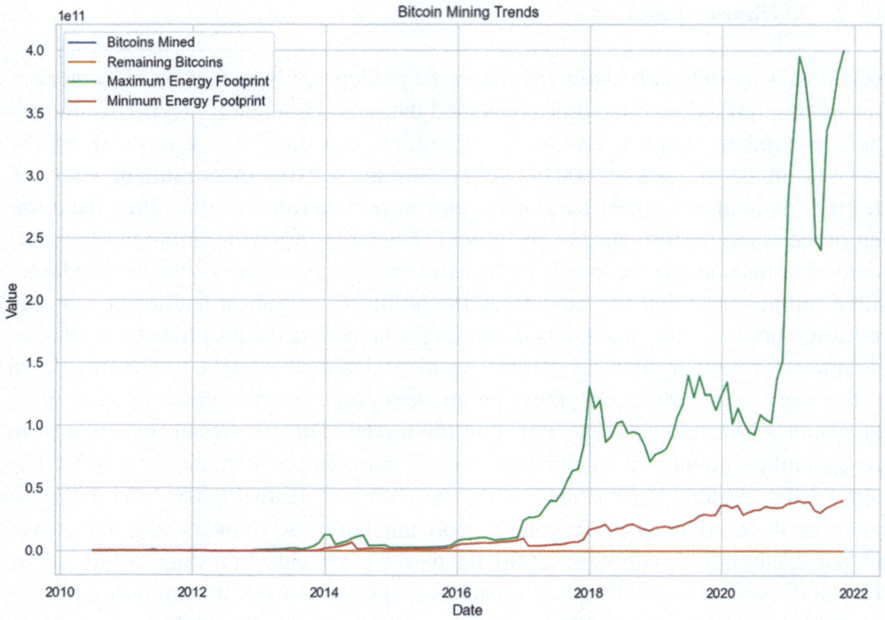

Fig. 3 Bitcoin mining trends (2010–2022): Analysis of mined vs. remaining Bitcoins and their correlation with maximum and minimum energy footprints

is sufficient for getting the optimal balance between the accuracy of the training and time being taken, while max_depth = 10 gives the number of levels that the trees can grow to, up to 10 levels, so that each tree is able to capture moderate complexity in the patterns. This limited depth helps control the formation of deep trees which could map noise in the data. A depth of 10 is used in the process of effective pattern recognition while overcoming the issues with excessive complexity and overfitting (Fig. 3).

4 Results

4.1 ARIMA Model with Dataset 1

The best_order = (5, 0, 5) was the optimal parameters for the ARIMA model with our dataset 1, this means that the 5 lag observations are used to predict the future values (visibility range), and there is no need to make difference to get a stationary result ($d = 0$), the last 5 lagged forecast errors are incorporated in the moving average component ($q = 5$). The computed Mean Absolute Error (MAE) for the model is 33,225,120,048.92, which indicates that the average of the differences between the forecasts and the actual values is about 33.2 billion units of energy consumption,

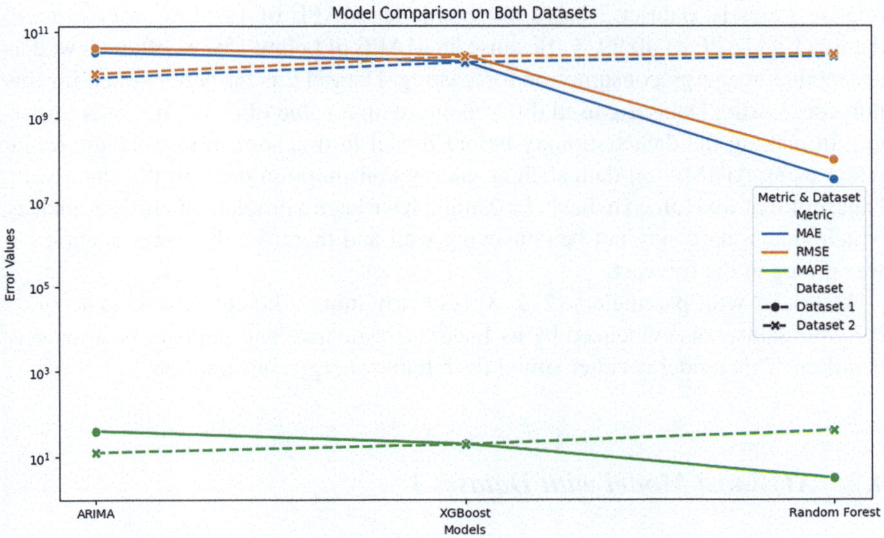

Fig. 4 Comparative analysis of ARIMA, XGBoost, and Random Forest models across two datasets using MAE, RMSE, and MAPE metrics

which is equal to the absolute value of prediction errors. A high level of the RMSE is denoted from 42,120,050,718.79, which depicts that there is a relatively large number of large absolute residuals in the model since RMSE gives a lot of consideration to the large errors. Lastly, Mean Absolute Percentage Error (MAPE) is calculated to be approximately 40.99% which reveals that the average forecast error is voluming approximately 41% of the actual value which gives the relative measure of the accuracy of the observed data as envisaged by the model that can be seen in the Figs. 1 and 4.

Although it can be deduced that the model makes some points about the data it analyzed, the amounts of forecast errors mean that there is still room for improvement; it is possible to refine the model, to analyze more features or try different models.

4.1.1 ARIMA Model with Dataset 2

The ARIMA model for Dataset 2, optimized with parameters ($p = 2$, $d = 2$, $q = 3$), produced the following results: Mean Absolute Error (MAE) of approximately $8,610,692,565,418, Root Mean Squared Error (RMSE) around $10,643,489,877,368 and Mean Absolute Percentage Error (MAPE) of about 12.81%. Error estimators also show that the result of Dataset 2 with parameters (2, 2, 3) outperforms the results of Dataset 1 with a nearly 74% lower MAE and RMSE. These small movements imply that dataset 1 has much fewer catastrophic errors and is more likely to predict actual values with less difference in ascertaining data patterns. Also, on relative grounds, dataset 2 has a much smaller MAPE of 12.81 % in contrast to dataset 1's MAPE of 40.99 %. Because an MAPE of below 15% is often viewed as acceptable in energy consumption forecasting, Dataset 2 is far better suited for this purpose. Lastly, Dataset 2 used differencing with a value of $d = 2$, this was important in making the data stationary before model fitting, something very important while using ARIMA on data such as energy consumption data. In the same vein, Dataset 1 that was forced to have $d = 0$ might have been a problem of non-stationarity which annual data may not be conveying well and therefore the lower accuracy it was giving in the forecast.

Dataset 2 with parameters (2, 2, 3) is clearly more efficient as well as accurate for your dataset, as evidenced by its lower error metrics and improved stationarity handling. This model is better suited for reliable energy consumption.

4.2 XGBoost Model with Dataset 1

For the given dataset, XGBoost model reached a total test MAE of approximately 19.99 * 10^9 that is on average 19.99 billion the model is off from the true value. The Root Mean Squared Error (RMSE) is around 28.25 billion; indicating that the present model undergoes considerably more damage from relatively large errors due to the squared nature of the RMSE. The MAPE is 20.21% therefore the model predictions are on average 20.21% off from the actual values. These metrics seem to be reasonable; nonetheless, they reveal some potential for further refinement specific to managing larger deviations. MAE and RMSE indicate the error margin between predicted and actual energy demand values on average and on their mean squared scale. In this case, both the values are high; it means that the gap between the predicted values and actual data must be quite large. As we can clearly see, the plot of RMSE is greater than that of MAE, making RMSE considerably larger, which means there is sensitivity to large errors in the dataset.

It is proposed that an MAPE of ~20.21% implies that on average your model makes mistakes of about 20% in its predictions from the actual values. Nevertheless, this gives a certain level of predictive ability; an even lower MAPE, ideally below 10–15%, for energy forecasting. Based on these outcomes, the model accurately estimates the large-scale consumption trends and might require tweaking, to increase its effectiveness, for exact prediction. The resulting XGBoost model was characterized by an average MAPE of 20% which is generally acceptable for most energy forecasting use-cases, yet, a bit imprecise compared to the high-precision models that might be required at times. Improving the performance might mean tuning the hyperparameters, using more features with or without transformations and or other categories of models.

4.2.1 XGBoost Model with Dataset 2

Evaluation of the model shows that the Mean Absolute Error (MAE) stood at around 19.09 billion implying that the models predictions were off by an average of 19.09 billion. The RMSE is about 26.21 billion, making it clear that the prediction errors of a larger magnitude are more damaging due to the use of squared term for RMSE. The last measure, the Mean Absolute Percentage Error (MAPE) is slightly bigger than the previous one and it indicates to us that the average error of the model corresponds to 19.69% of the real values. These values indicate better accuracy where large errors are less than the previous results obtained.

4.2.2 Comparison

The new XGBoost model shows an improvement, with both lower MAE and RMSE (MAE: Decreased from ~19.99 billion to ~19.09 billion for MSE and from ~28.25 billion to ~26.21 billion for RMSE). These reductions of MAE and RMSE suggest that the new model is generally making smaller of average errors and less large deviations. MAPE has also reduced from ~20.21% to ~19.69%. These reductions demonstrate that the new model is relatively more accurate than the actual values in most cases. The new XGBoost model's general performance looks to have gotten slightly better across all three evaluation metrics (MAE, RMSE, and MAPE) implying that tweaks were made to the model; perhaps in hyperparameters or the features, made the model more accurate. Nevertheless, the differences are neither so big nor significant. Although the new model brings some generalization improvement compared to previous one, there is still a space for tuning or trying to engineer new features that may bring better results.

4.3 Random Forest Model

4.3.1 Random Forest Model with Dataset 1

The Random Forest model worked really well for Dataset 1. We had a Mean Absolute Error (MAE) of roughly 34.78 million, so on average the model is way off by approximately 35 million units. As evident from this relatively small error margin, the model thus effectively predicts. The MAE was recorded at 96.796 million; however, the Root Mean Squared Error (RMSE) fell to 100.145 million. This means that most predictions are correct, but that there were some bigger ones with not much distortion across the line, making them more expensive for RMSE to ignore. In spite of that, the RMSE still points to a sufficient model performance for energy consumption data which can possess high values. With a Mean Absolute Percentage Error (MAPE) of 3.20%, the model has predicted its values, on average, to be off by about 3.2%. This is a fine result, given how often energy consumption values vary here, and has the added benefit of being able to forecast them.

4.3.2 Random Forest Model with Dataset 2

On the contrary, the performance of the Random Forest model with Dataset 2 was not as good. Dataset 2 had an MAE of 796.7 million which was a much greater amount than Dataset 1 indicating that the model's predictions are much less accurate to the actual values. This shows that the model has a greater MAE because it has more tasks in which to accurately predict energy consumption at this dataset. Even more so, the RMSE for Dataset 2 was 1.35 billion, much higher than the RMSE for Dataset 1. The fact that large errors drive the prediction for Dataset 2, and thereby a less reliable model performance, produces this substantial increase. Dataset 2's MAPE, or mean absolute percentage error, was 7.99%, meaning the model was 8 percent off on average, which is still okay, but not as good as we saw with Dataset 1.

4.3.3 Comparison with Other Models

Random Forest model performs clearly better than other models (XGBoost, ARIMA, Prophet), in Dataset 1 and Dataset 2. The MAE observed for Random Forest on Dataset 1 of 34.78 M is overwhelmingly better than the 19.99 B MAE for XGBoost, thus making Random Forest a much more reliable model for this dataset. Likewise, the quality of the MAPE for Random Forest of 3.20% far outperforms XGBoost's 20.21% that is Random Forest captures the underlying patterns in a much more visible way. For Dataset 2, Random Forest outperforms XGBoost, with MAE = 796.7 million vs. 19.09 billion for XGBoost. Moreover, Random Forest has a better MAPE than XGBoost (19.69%) and so its performance is also very good. Random Forest model performs way better than ARIMA. Random Forest results with MAE of 34.78 million is way too superior over around 19.99 billion MAE of ARIMA for Dataset 1 to estimate how Random Forest does better in forecasting energy consumption than ARIMA. Random Forest's MAPE of just 3.20% which is also lower than ARIMA 20.21% indicates better accuracy. Similarly, ARIMA is outperformed by Random Forest on Dataset 2 both in terms of lower error metrics for both MAE and MAPE. Further, in comparison with Prophet, Random Forest exhibits a striking performance leap. The MAE score of Prophet is 26.59 billion for Dataset 1 but Random Forest scores only 34.78 million, while the MAPE scores 41.47% whereas Prophet scores 3.20%. The Random Forest performs significantly better, and even for the Dataset 2 Prophet drops behind Random Forest significantly on MAE and MAPE.

Overall however, the Random Forest model performs consistently well for both datasets especially with Dataset 1, having a low MAE, RMSE, MAPE. Its performance on Dataset 2 wasn't as cutting edge though (unless I've missed a wildcard) but even still it performed better than the other models tested including a rather popular XGBoost, ARIMA, and Prophet. Since the Random Forest model is able to efficiently represent the inherent trend within the data, the model renders well for energy consumption prediction and its K-Fold cross-validation error is contoured to outperform other models with respect to accuracy and other error metrics.

4.4 Prophet

Prophet is one open source tool that has been made available to the public by Facebook for the production of good forecasts from time series data, namely the seasonal characteristic and from data of several seasons in this company. Thanks to it, a consumer can easily incorporate holidays and other seasonality into the system starting from the present moment. However, while applying the Prophet Model we arrived at approximately 26,585 million of MAE, approximately 31,033 million of RMSE and 41.47% of MAPE. These metrics show the actual values of these predicted price values are apprehended and quite distorted and hence are not as effective in this sense as the other models like "Random Forest" and "XGBoost" which have made fairly low rate of errors. This indicates that there is an opportunity to decrease the error or that other methods of modeling should be used to develop sound estimations of energy consumption.

4.5 CatBoost

Energy consumption prediction using the CatBoost model is as follows: MAE: 1.80 billion units, RMSE: 3.22 billion units, MAPE: 96.19%. The high error rates mean that the model is off by an average of nearly 96 percentage points when making predictions of the test set values against the predicted values. This seems remarkably bad when compared with other models which have been tried before, including Random Forest, which has an MAPE of 7.99%. Amidst the outcomes, it can be concluded that CatBoost is not helpful in this dataset: the further investigation of the data quality and losing, feature engineering and, potentially, the models' specifications is essential to improving the forecast's precision. Comparison of errors across MAE, RMSE, and MAPE of different models is depicted in Fig. 4. Detailed results are shown below.

5 Discussion

This chapter aims to identify and solve the ever increasingly growing energy consumption in blockchain mining with an accent on cryptocurrencies like Bitcoin. The increased pace of the adoption and increase of blockchain technology has heightened concerns over the environmental and economic impact of the energy expensive Proof-of-Work (PoW) consensus mechanism [5, 6, 30]. How much Bitcoin and other cryptocurrencies actually consume in energy, and how this changes over time, is the focus of this chapter. The chapter examines historical trends and compares the annual energy consumption of various countries and draws comparisons of the potential long-term implications if current mining practices continue. One of the key goals of this research is to use highly developed machine learning approaches, such as ARIMA, XGBoost, and Random Forest to predict future energy usage scenarios. The first group of these models takes into account different parameters like mining difficulty, hardware evolution, and trend-related market models that are used to make accurate predictions. The intent of these predictions is to inform the strategy of reducing energy usage in blockchain mining operations [31, 32]. This study is also advocating for renewable energy sources integration in the mining operations for cutting down on fossil fuel use and carbon emissions. The paper also reports how blockchain-based energy management systems such as smart grids and peer-to-peer trading can be so helpful to better manage the sustainability and efficiency of energy networks [33]. In the end, the main contribution of this chapter is to provide guidance to the global debate towards creating sustainable blockchain ecosystems through providing actionable solutions for achieving energy efficiency. By means of analysis and recommendations, the study aims to harmonize blockchain technology with diverse environmental and sustainable development targets so that it remains viable and makes a positive contribution for the future. Also, in this section, we reflect on the results of various models taken together, particularly for efficacy in predicting the energy consumption with two different datasets. MAE, RMSE, and MAPE—evaluation metrics were used to evaluate the accuracy of each model. On Dataset 1, ARIMA (5, 0, 5) had high forecast errors with an MAE of 33.2 billion and RMSE of 42.12 billion and MAPE of 40.99%. The high error values show that ARIMA had a hard time capturing the complexity of the energy consumption data in this dataset. Though in Dataset 2 parameters (2, 2, 3) yielded an MAE of 8.6 billion, RMSE of 10.64 billion, and MAPE of 12.81%, the MAE, RMSE, and MAPE were improved using ARIMA. Nevertheless, the model's performance was better, although it was not optimal in fact for this task, and in particular handles non-stationary time series data. For Dataset 1, XGBoost takes an MAE of 19.99 billion, RMSE of 28.25 billion, and an MAPE of 20.21% that outperforms ARIMA and yet still have noticeable errors in RMSE and MAPE. This makes XGBoost a very good model, but it needs more work on feature engineering and hyperparameters tuning to extract out the patterns of energy consumption. Dataset 2 was slightly improved by XGBoost with an MAE = 19.09 billion, RMSE = 26.21 billion, and an MAPE = 19.69%. As before, these incremental improvements indicate how XGBoost might further optimize, but still could not outperform Random Forest. Dataset 1, however, was greatly outperformed by Random Forest compared to both ARIMA and XGBoost, with MAE and RMSE of 34.78 million and 100.145 million, respectively, and MAPE of 3.20%. These results demonstrate the Random Forest's effectiveness in working with complex, high-dimensional data and in accurate prediction of energy consumption patterns. On Dataset 2, the MAE of the model was 796.7 million, RMSE of 1.35 billion, and MAPE of 7.99% (slightly worse than ARIMA and XGBoost in accuracy), this is an improvement over the results for Dataset 1. On both datasets, the model's capability to capture energy consumption variations is demonstrated and its performance.

6 Conclusion

This paper aims to address the challenges of accurately forecasting energy consumption in the context of Bitcoin mining using advanced machine learning models. For this, we worked upon two distinct datasets, enabling a robust comparative analysis of models such as ARIMA, XGBoost, Random Forest, Prophet, and CatBoost. Evaluation demonstrated that Random Forest emerged as the most reliable model, with the lowest MAPE, affirming its suitability for this task. Our studies explore Bitcoin's energy footprints; the majority are theoretical and lack empirical validation. This paper aims at fine lining the gap by offering a data-driven, practical approach to energy consumption prediction. This paper provides actionable insights into the minutes of Bitcoin's energy demands and lays the foundation for future research, emphasizing the scalability and real-world applicability of machine learning in addressing cryptocurrency-related energy challenges. In this study, Random Forest was the best choice for energy consumption prediction for the two datasets because it had the lower MAE, RMSE, and MAPE over ARIMA and XGBoost. That is, its superior performance comes from its ability to model complex relationships and interactions between the data, whereas the non-stationary time series data caused the ARIMA to perform very badly, in terms of their predictive accuracy. XGBoost had promise, but couldn't quite achieve the same precision as Random Forest, requiring a little more refinement to achieve its level of performance. Random Forest's robustness towards a large and complex dataset secures its position as the most trustable model for predicting energy consumption in blockchain mining. We further tried to improve the ARIMA and XGBoost by both tuning hyperparameters and feature engineering, but they could not solve this specific task sufficiently. This raises the possibility of future research to enhance these models making them viable alternatives. Random Forest is the best choice to solve energy prediction issues in mining with cryptocurrency due to its accuracy and robustness, and the study shows the necessity of exploring the future of other applicable models to increase accuracy.

References

1. Bajra, U. Q., Rogova, E., & Avdiaj, S. (2024). Cryptocurrency blockchain and its carbon footprint: Anticipating future challenges. *Technology in Society, 77*, 102571.
2. Kohli, V., Chakravarty, S., Chamola, V., Sangwan, K. S., & Zeadally, S. (2023). An analysis of energy consumption and carbon footprints of cryptocurrencies and possible solutions. *Digital Communications and Networks, 9*(1), 79–89.
3. Qiu, J., Zhao, J., Wen, F., Zhao, J., Gao, C., Zhou, Y., Tao, Y., & Lai, S. (2023). Challenges and pathways of low-carbon oriented energy transition and power system planning strategy: A review. *IEEE Transactions on Network Science and Engineering., 11*(6), 5396–5416.
4. Singh, I., Kaur, A., Agarwal, P., & Idrees, S. M. (2024). Enhancing security and transparency in online voting through blockchain decentralization. *SN Computer Science, 5*(7), 921.

5. Ahmad, T., Madonski, R., Zhang, D., Huang, C., & Mujeeb, A. (2022). Data-driven probabilistic machine learning in sustainable smart energy/smart energy systems: Key developments, challenges, and future research opportunities in the context of smart grid paradigm. *Renewable and Sustainable Energy Reviews, 160*, 112128.
6. Bada, A. O., Damianou, A., Angelopoulos, C. M., & Katos, V. (2021, July). Towards a green blockchain: A review of consensus mechanisms and their energy consumption. In *In 2021 17th international conference on distributed computing in sensor systems (DCOSS)* (pp. 503–511). IEEE.
7. Wongthongtham, P., Marrable, D., Abu-Salih, B., Liu, X., & Morrison, G. (2021). Blockchain-enabled Peer-to-Peer energy trading. *Computers & Electrical Engineering, 94*, 107299.
8. Kanika, Chakraverty, S., Chakraborty, P., Aggarwal, A., Madan, M., & Gupta, G. (2022). Enriching WordNet with subject specific out of vocabulary terms using existing ontology. In *Data Engineering for Smart Systems: Proceedings of SSIC 2021* (pp. 205–212). Springer Singapore.
9. Wang, T., Hua, H., Wei, Z., & Cao, J. (2022). Challenges of blockchain in new generation energy systems and future outlooks. *International Journal of Electrical Power & Energy Systems, 135*, 107499.
10. Sahebi, I. G., Mosayebi, A., Masoomi, B., & Marandi, F. (2022). Modeling the enablers for blockchain technology adoption in the renewable energy supply chain. *Technology in Society, 68*, 101871.
11. Kumari, A., Gupta, R., Tanwar, S., & Kumar, N. (2020). Blockchain and AI amalgamation for energy cloud management: Challenges, solutions, and future directions. *Journal of Parallel and Distributed Computing, 143*, 148–166.
12. Luo, X., & Mahdjoubi, L. (2024). Towards a blockchain and machine learning-based framework for decentralized energy management. *Energy and Buildings, 303*, 113757.
13. Jin, W., Li, C., & Zheng, M. Y. (2024). Sustainable energy management in electric vehicle secure monitoring and blockchain machine learning model. *Computers and Electrical Engineering, 115*, 109093.
14. Nykyri, M., Kärkkäinen, T. J., Levikari, S., Honkapuro, S., Annala, S., & Silventoinen, P. (2022). Blockchain-based balance settlement ledger for energy communities in open electricity markets. *Energy, 253*, 124180.
15. Gill, S. S., Tuli, S., Xu, M., Singh, I., Singh, K. V., Lindsay, D., Tuli, S., Smirnova, D., Singh, M., Jain, U., & Pervaiz, H. (2019). Transformative effects of IoT, blockchain and artificial intelligence on cloud computing: Evolution, vision, trends and open challenges. *Internet of Things, 8*, 100118.
16. Andoni, M., Robu, V., Flynn, D., Abram, S., Geach, D., Jenkins, D., McCallum, P., & Peacock, A. (2019). Blockchain technology in the energy sector: A systematic review of challenges and opportunities. *Renewable and Sustainable Energy Reviews, 100*, 143–174.
17. Ma, C. Q., Lei, Y. T., Ren, Y. S., Chen, X. Q., Wang, Y. R., & Narayan, S. (2024). Systematic analysis of the blockchain in the energy sector: Trends, issues, and future directions. *Telecommunications Policy, 48*(2), 102677.
18. Zhao, M., Liao, L., Liang, P., Li, M., Dong, X., Yang, Y., Wang, H., & Zhang, Z. (2024). Energy blockchain in smart communities: Towards affordable clean energy supply for the built environment. *Energy Engineering, 121*(8), 2313–2330.
19. Owusu, P. A., & Sarkodie, S. A. (2022). Dataset on bitcoin carbon footprint and energy consumption. *Dataset, 42*, 108252. https://doi.org/10.6084/m9.figshare.19442933.v1
20. Zhang, G. P. (2003). Time series forecasting using a hybrid ARIMA and neural network model. *Neurocomputing, 50*, 159–175.
21. Shi, L., Qian, C., & Guo, F. (2022). Real-time driving risk assessment using deep learning with XGBoost. *Accident Analysis & Prevention, 178*, 106836.
22. Ghosh, D., & Cabrera, J. (2021). Enriched random forest for high dimensional genomic data. *IEEE/ACM Transactions on Computational Biology and Bioinformatics, 19*(5), 2817–2828.

23. ArunKumar, K. E., Kalaga, D. V., Kumar, C. M. S., Chilkoor, G., Kawaji, M., & Brenza, T. M. (2021). Forecasting the dynamics of cumulative COVID-19 cases (confirmed, recovered and deaths) for top-16 countries using statistical machine learning models: auto-regressive integrated moving average (ARIMA) and seasonal auto-regressive integrated moving average (SARIMA). *Applied Soft Computing, 103*, 107161.
24. Khashei, M., & Bijari, M. (2010). An artificial neural network (p, d, q) model for timeseries forecasting. *Expert Systems with Applications, 37*(1), 479–489.
25. Wen, Z., Zhou, R., & Su, H. (2022). MR and stacked GRUs neural network combined model and its application for deformation prediction of concrete dam. *Expert Systems with Applications, 201*, 117272.
26. Dash, P. K., Satpathy, H. P., & Liew, A. C. (1998). A real-time short-term peak and average load forecasting system using a self-organising fuzzy neural network. *Engineering Applications of Artificial Intelligence, 11*(2), 307–316.
27. Idrees, S. M., Alam, M. A., Agarwal, P., & Ansari, L. (2019). Effective predictive analytics and modeling based on historical data. In *Advances in Computing and Data Sciences: Third International Conference, ICACDS 2019, Ghaziabad, India, April 12–13, 2019, Revised Selected Papers, Part II 3* (pp. 552–564). Springer Singapore.
28. Li, J., Liu, X., & Shao, X. (2024). Collaborative carbon emission reduction in power supply and demand entities based on blockchain technology. *International Journal of Electrical Power & Energy Systems, 157*, 109840.
29. Teufel, B., Sentic, A., & Barmet, M. (2019). Blockchain energy: Blockchain in future energy systems. *Journal of Electronic Science and Technology, 17*(4), 100011.
30. Wang, L., Jiang, S., Shi, Y., Du, X., Xiao, Y., Ma, Y., Yi, X., Zhang, Y., & Li, M. (2023). Blockchain-based dynamic energy management mode for distributed energy system with high penetration of renewable energy. *International Journal of Electrical Power & Energy Systems, 148*, 108933.
31. Boumaiza, A. (2024). A blockchain-centric P2P trading framework incorporating carbon and energy trades. *Energy Strategy Reviews, 54*, 101466.
32. Shahjahan Alam, K., Daiyan Kaif, A. M. A., & Das, S. K. (2024). A blockchain-based optimal peer-to-peer energy trading framework for decentralized energy management with in a virtual power plant: lab scale studies and large scale proposal. *Applied Energy, 365*, 123243.
33. Madhwal, Y., Yanovich, Y., Coveri, M., & Marina, N. (2024). Blockchain-enhanced hydrogen fuel production and distribution for sustainable energy management. *Blockchain: Research and Applications, 5*(4), 100229.

Smart Farming Practices for Reinventing Agriculture Using Artificial Intelligence Techniques

Sheetal Mavi, Neetu Mittal, and Pradeepta Kumar Sarangi

1 Introduction

Smart farming is a new strategy that leverages cutting-edge technologies like the Internet of Things (IoT) and Artificial Intelligence (AI) to revolutionize agriculture. Thanks to technology like the Internet of Things (IoT) and Artificial Intelligence (AI), modern farming practices are changing dramatically. Farmers are interested because it contributes to more sustainable and effective farming. AI is a great help to agriculture. It ensures that we make efficient use of our resources and aids in addressing issues brought on by climate change. IoT, on the other hand, focuses on connecting and intelligently utilizing farm equipment. As a result, farms will be able to gather and utilize real-time data to inform smarter decisions. A number of aspects related to farming, including fertilizer use, animal welfare, crop management, soil health, weather forecasting, insect control, water management, crop monitoring, machinery optimization, and field operations, are heavily reliant on the Internet of Things and artificial intelligence. Government funds and company incubators are bolstering these concepts, and initiatives like "AI for all" are establishing guidelines for the application of these technologies in agriculture. Encouraging agricultural technology companies to expand and improve farming is the aim. In this chapter, we will discuss the concept of Integration of Smart Farming, Artificial Intelligence, and Internet of Things.

S. Mavi (✉) · N. Mittal
Amity University, Noida, Uttar Pradesh, India
e-mail: nmittal1@amity.edu

P. K. Sarangi
Chitkara University School of Engineering and Technology, Chitkara University, Baddi, Himachal Pradesh, India

S. M. Idrees et al. (eds.), *Blockchain and Machine Learning Innovations*, Signals and Communication Technology,
https://doi.org/10.1007/978-3-031-88095-7_4

2 The Concept of Smart Farming

With the integration of Internet of Things (IoT) and Artificial Intelligence (AI), an innovative agricultural approach smart farming, integrates advanced technologies to optimize farming processes. By employing sensors and interconnected IoT devices enables real-time monitoring of crucial factors like soil moisture, crop health, and environmental conditions. With the help of AI processes, the data gets processed and provides actionable insights, enabling farmers to make informed and informed decisions. This allows targeted resource application and reduces waste which enhances agriculture and contributes to Precision Agriculture. Remote management can be facilitated by farm connectivity, automated machinery, and robotics streamline tasks. Smart farming is a paradigm of sustainability and efficiency that is transforming conventional methods for increased output. With the help of diagram shown in Fig. 1, we tried to explain the concept of smart farming. The Internet of Things devices produce the data. Artificial intelligence is used to process this data, which helps with precision agriculture since the processed data becomes informed data which ultimately leads to sustainability.

3 The Role of Blockchain in Enhancing Smart Agriculture

Blockchain technology is revolutionizing the agricultural sector by providing innovative solutions that enhance traceability, transparency, and efficiency throughout the supply chain [9–11]. Table 1 summarizes key aspects of blockchain applications in agriculture, including their descriptions, practical examples, benefits, and references for further reading. By leveraging blockchain, stakeholders can improve food safety, reduce fraud, automate transactions, and promote sustainable practices, ultimately leading to a more resilient and trustworthy agricultural ecosystem.

4 The Impact of Artificial Intelligence on Smart Agriculture

Artificial Intelligence refers to the technology which is incorporated in machines to give them the ability to make choices without human input. AI has changed the way we practice agriculture with its amazing and enhanced applications. It has been observed from past 50 years that AI is helpful in many fields. In Agriculture, from planting seeds to harvesting crops, AI is helping farmers with numerous of challenges. For keeping crops protected from illnesses, managing storage properly, limiting the use of pesticides, efficiently controlling weeds, and enhancing irrigation and drainage systems, AI is playing a major role in resolving these issues that are critical to farming's overall viability. With the help of AI, farming has become more effective and robust in addressing these problems. The implementation of AI is

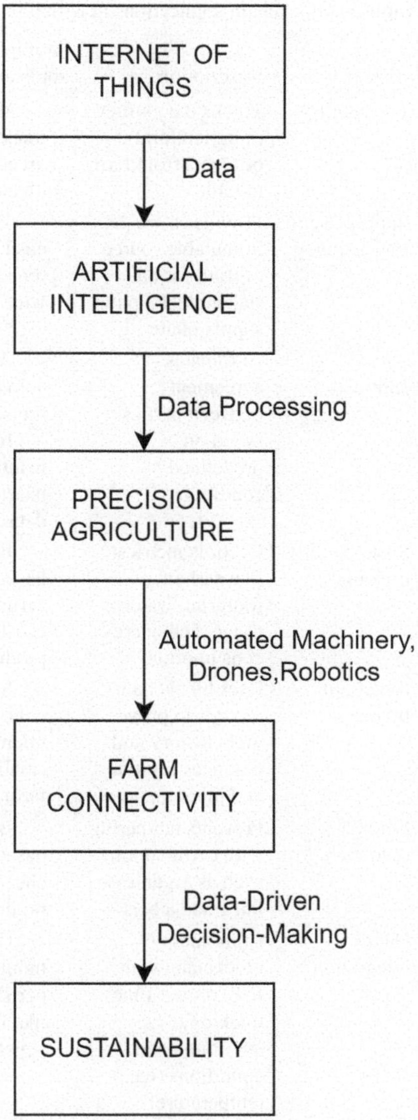

Fig. 1 Concept of smart farming

increasing day by day in the agriculture industry to assist farmers in raising yields, decreasing input costs, and improving overall operational efficiency. This technology can improve soil management, automate crop management, monitor and identify issues with crops. AI is capable of making judgments and identifying patterns from the data.

Artificial Intelligence (AI) has been used in precision farming to maximize the crop yield by proper management of fertilizers, nutrients, water supply, and early detection of diseases in plants. AI can help the farmers to improve the quality of

Table 1 Blockchain applications in agriculture: Key aspects, examples, and benefits

Aspect	Description	Examples/ applications	Benefits
Traceability	Tracks the journey of agricultural products from farm to table	– Walmart tracking leafy greens using IBM blockchain	– Ensures food safety – Builds consumer trust – Quick source identification during recalls
Supply chain transparency	Provides a single, immutable source of truth for all stakeholders in the supply chain	– VeChain ensuring transparency in wine supply chains	– Reduces fraud – Optimizes logistics – Enhances collaboration across stakeholders
Smart contracts	Automates agreements between parties based on predefined conditions	– AgriDigital automating grain transactions – IoT-triggered insurance payouts based on weather data	– Ensures timely payments – Reduces reliance on intermediaries – Improves risk mitigation
Sustainability tracking	Records metrics like carbon footprint, water usage, and energy consumption	– Blockchain-based sustainability certifications for eco-friendly products	– Appeals to environmentally conscious consumers – Encourages sustainable practices
Access to finance	Uses blockchain records to prove yield history and practices for loans and investments	– Smart contract-enabled microloans to smallholder farmers	– Facilitates financial inclusion – Reduces loan approval time – Enhances farmer trustworthiness
Fraud reduction	Prevents tampering with certifications such as organic or fair trade labels	– Blockchain-based authenticity checks for organic produce	– Ensures genuineness of products – Reduces counterfeiting
Data integration	Combines blockchain with IoT for real-time tracking of environmental conditions (e.g., temperature)	– Cold-chain monitoring for perishable goods like fruits and vegetables	– Improves product quality during transportation – Enhances decision-making based on real-time data

their harvests and also cut the cost. Through the implementation of neural networks in deep learning, in which training on big datasets like categorizing and detecting objects in photographs, machines may learn to identify patterns and further make predictions on the basis of insights [1]. As the population grows, so does the need for agriculture, and artificial intelligence (AI) is emerging as a critical tool for quickly and accurately spotting trends so that businesses may leverage linked data. Through the application of AI and machine learning, digital data may be utilized to assess agricultural accuracy, enabling accurate, and remote monitoring. Technology utilizing artificial intelligence (AI) can alleviate the labor shortage problem while increasing productivity and efficiency [2]. Natural language processing, computer vision, speech recognition, pattern recognition, and machine learning are all subsets that are included in the set AI cycle. AI technologies can create a digital brain by using complex algorithms through directed learning, unsupervised learning, and reinforcement learning. The future of Smart Farming relies on the adoption of AI technologies to increase efficiency, productivity, and sustainability. Machine learning and AI have shown promising results in other countries, and the agriculture industry can benefit from their implementation [3].

These challenges encompass:

1. Pervasive occurrences of crop diseases
2. Deficiencies in the management of storage facilities
3. Rigorous control of pesticide applications
4. Systematic weed management
5. Inadequacies in both irrigation and drainage infrastructure [1].

The Internet of Things (IoT) is a smart and exciting technology that brings cool and useful solutions. It has been implemented in different areas like smart cities, traffic control, homes, healthcare, and even smart farming. In farming, IoT has made big progress in how we manage things. It helps connect all the devices and tools on the farm so they can work together to make smart decisions about things like watering crops and supplying fertilizer. As highlighted in recent studies [5], the application of IoT in agriculture is transforming traditional farming practices. In Fig. 2, we illustrate the synergy between the Internet of Things (IoT) and smart farming through the combination of various technologies.

The Internet of Things (IoT) is a force for good in all fields, changing people's lives by bringing intelligence to everything. This is a self-configuring network made up of several devices. The application of IoT to smart farming is revolutionizing

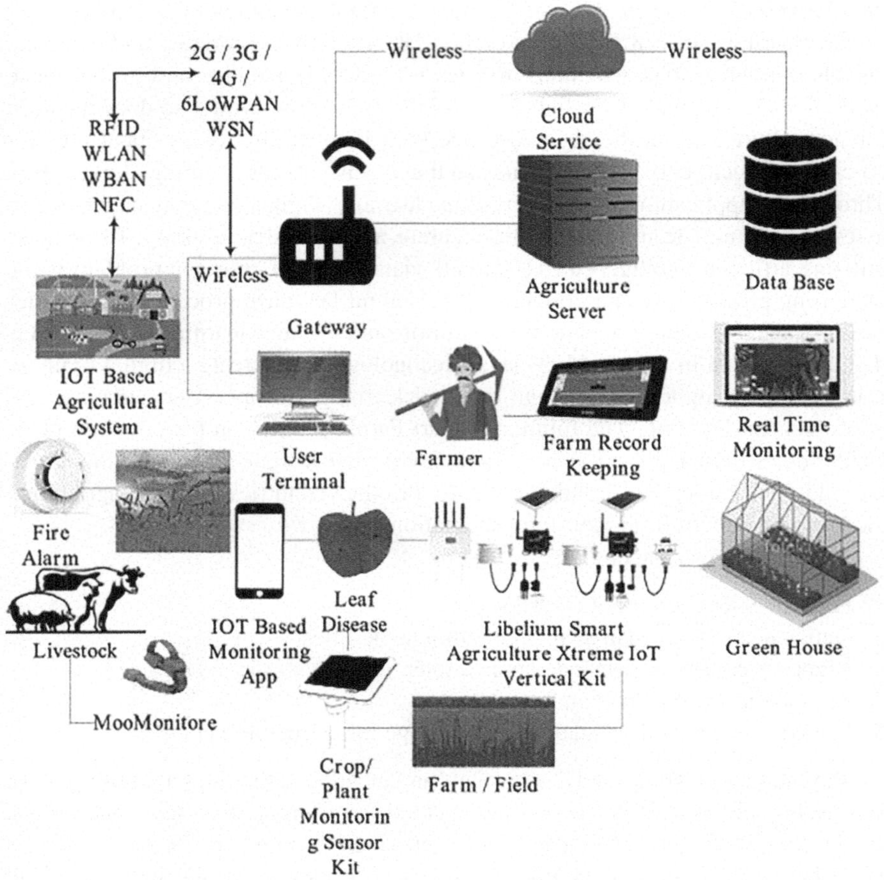

Fig. 2 Integrating technologies for agricultural advancement

traditional agricultural practices by decreasing waste from crops and boosting productivity while also saving farmers money.

Using the Internet of Things (IoT) in farming has changed the way we do things on the farm. It helps farmers make decisions based on data, making farming more efficient, cheaper, and better for the environment. So, countries that depend a lot on farming for their economy should think about using IoT in their farming methods [2]. In Fig. 3, we depict the implementation and application of smart farming across various scenarios utilizing Internet of Things (IoT) technologies.

The smart systems enhance the accuracy and efficiency of devices that raise livestock, monitor plant growth, and even help in effective crop yield production. To achieve this, wireless sensor networks (WSNs) are used to collect data from different sensing devices. It is essential that with IoT cloud services are also integrated to analyze and process the remote data that facilitates decision-making to implement the best data-driven decisions. Smart farm management requires ground sensors,

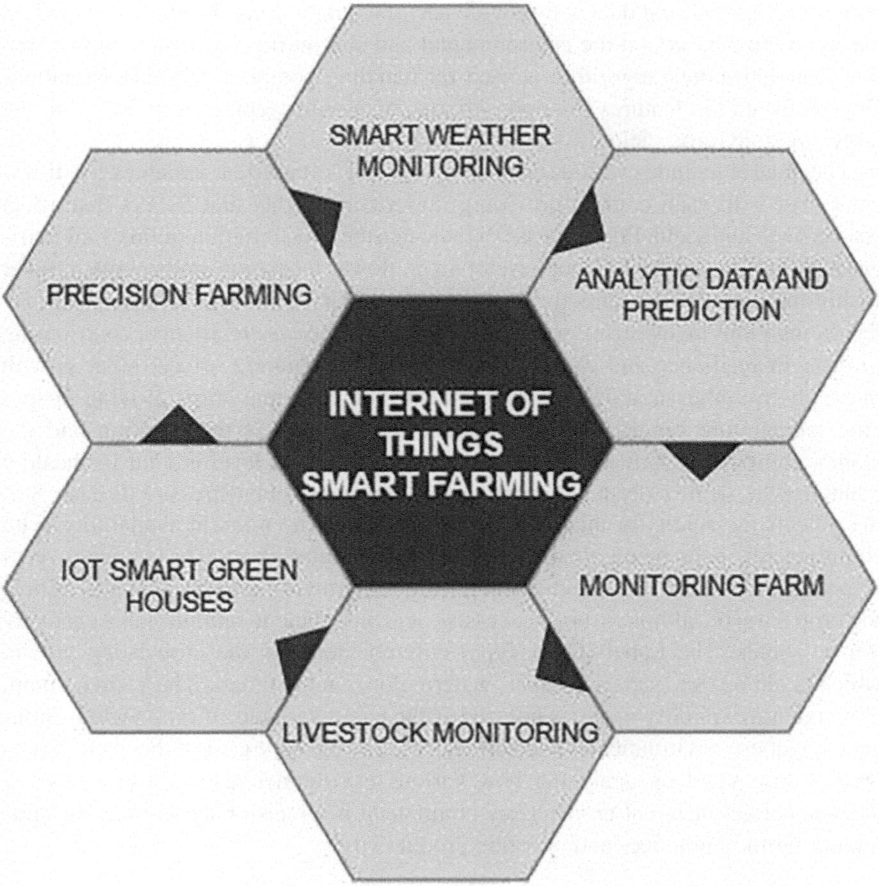

Fig. 3 IoT integration across diverse scenarios

ICT tools, control systems installed on robots, autonomous vehicles, and other automated devices. High-speed internet, advanced mobile devices, and satellites (to provide images and positioning) played a major role in the success of smart systems.

5 Implementing the Machine Learning Model for Crop Yield Prediction

In this section, the implementation of machine learning model has been done which predicts crop yields based on environmental factors and soil nutrients. Predicting crop yields is a critical task in agriculture, as it helps farmers make informed decisions about resource management, irrigation, fertilization, and planting strategies. Machine learning models, especially regression models, can identify complex patterns in agricultural data and provide accurate predictions. To predict crop yield based on the dataset and the environmental and soil nutrient variables. In this section, Random Forest algorithm is used for handling complex, non-linear relationships between the features (like phosphorus, potassium, temperature, etc.) and the target variable (crop yield).

The dataset includes various environmental and soil nutrient variables for different crops, with each column providing important insights into factors that affect crop growth and yield. Phosphorus (P) indicates the concentration of this vital nutrient in the soil, essential for root development, flower formation, and energy transfer within the plant, which directly impacts crop yields. Potassium (K) is crucial for regulating plant metabolism, water absorption, and disease resistance, contributing to drought resilience and overall plant health. Temperature affects plant growth rates, photosynthesis, and crop development, with different crops thriving in specific temperature ranges. Humidity influences the rate of transpiration and the plant's ability to absorb water, and maintaining the right level is vital for healthy plant growth, while extreme humidity levels may cause plant stress or disease. Soil pH reflects the acidity or alkalinity of the soil, affecting nutrient availability, with slightly acidic soils being ideal for most plants. Rainfall is another key factor, representing the amount of precipitation during the growing season, which is critical for crop growth, although both excessive and insufficient rainfall can negatively impact yields. The Label (Crop Type) column identifies the crop being grown, which could be rice, papaya, grapes, watermelon, among others. The Yield column, although not explicitly given, is inferred as the target variable, as crop yield is influenced by these environmental and soil factors. The primary goal of this dataset is to predict crop yield by analyzing how various environmental conditions and soil nutrients affect different crops, a key component of precision agriculture for optimizing farming practices and boosting productivity.

5.1 Feature Selection

The independent variables (features) that influence the crop yield prediction are explained in Table 2.

Table 2 Key factors influencing crop growth and development

P (phosphorus)	Nutrient that affects root development and energy transfer
K (potassium)	Affects water absorption, metabolism, and disease resistance
Temperature	Influences plant growth and development
Humidity	Affects transpiration and water uptake
pH	Influences nutrient availability in the soil
Rainfall	Affects water supply to the crop during the growing season

5.2 Model Evaluation

The model's performance is evaluated using:

(i) Mean Squared Error (MSE): Measures the average squared difference between predicted and actual values. A lower value indicates better performance.
(ii) R-squared (R^2): Represents the proportion of variance explained by the model. Values closer to 1 indicate a better fit.

The target variable in this case is the crop type, which is encoded as a numeric value. Ideally, we would predict actual crop yields, but for this example, crop type serves as a proxy for yield and is predicted based on various environmental and soil factors. The dataset is divided into training and testing sets, with 80% allocated for training the model and 20% used for testing its performance. This ensures that the model is trained on one subset of the data and evaluated on another, promoting better generalization to unseen data. For model training, a Random Forest Regressor is employed, an ensemble learning method that combines multiple decision trees to make predictions. The model is trained using the training data (X_train, y_train). Fig. 4 displays three scatter plots illustrating the relationship between crop yield and the application of three essential nutrients: Nitrogen (N), Phosphorus (P), and Potassium (K). Each plot shows the yield (on the vertical axis) against the corresponding nutrient level (on the horizontal axis).

(i) **Nitrogen (N):** The scatter plot for Nitrogen (N) shows a clear positive correlation between N application and crop yield. As the N level increases, the yield tends to increase as well. However, there seems to be a plateauing effect, where further increases in N beyond a certain point do not lead to significant yield gains.
(ii) **Phosphorus (P):** The scatter plot for Phosphorus (P) exhibits a less distinct relationship compared to Nitrogen. While there is a slight positive trend, the data points are more scattered, indicating that the impact of P on yield is less consistent.
(iii) **Potassium (K):** The scatter plot for Potassium (K) reveals a more complex pattern. There appears to be a narrow range of K levels where the yield is highest. Both below and above this optimal range, the yield decreases. This suggests that K application should be carefully managed to avoid both deficiency and over-application.

The plot in Fig. 5 depicts the relationship between crop yield and rainfall. Each point represents a data point with its x-coordinate indicating the rainfall level and its y-coordinate showing the corresponding yield. There appears to be a positive

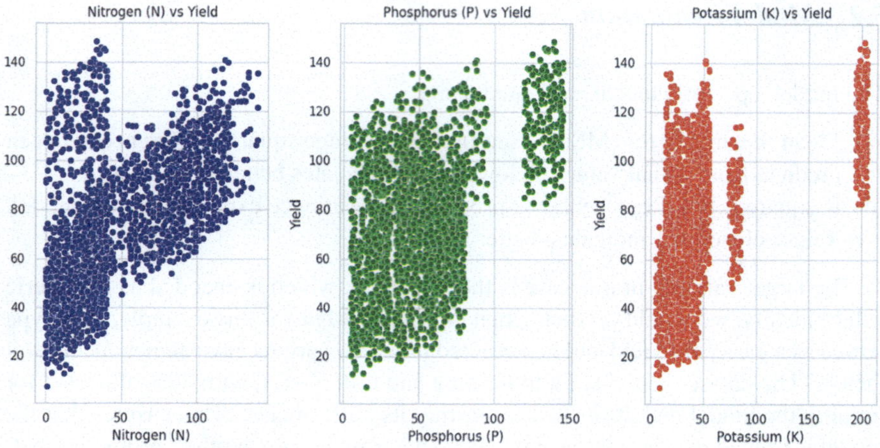

Fig. 4 Impact of nitrogen, phosphorus, and potassium on crop yield

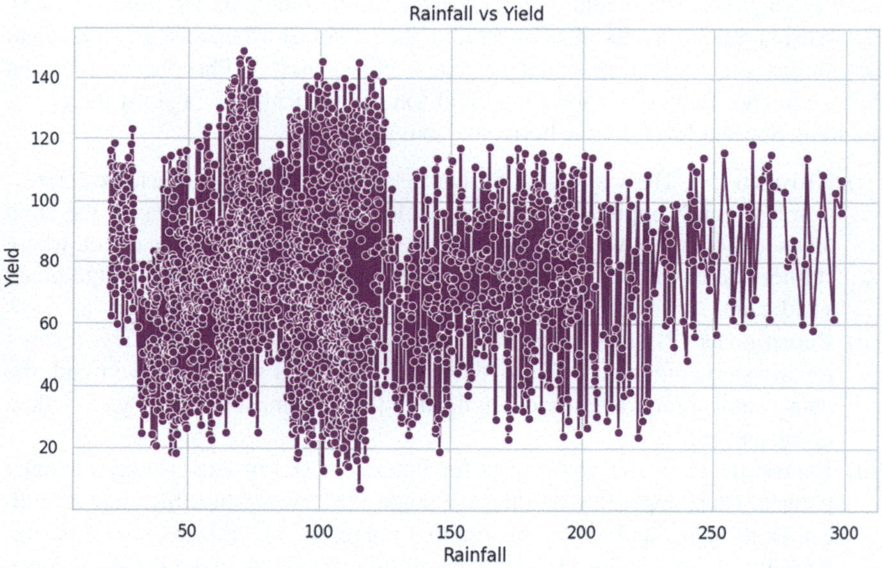

Fig. 5 Relationship between rainfall and crop yield

correlation between rainfall and yield. As rainfall increases, the yield also tends to increase. This is indicated by the general upward trend in the data points.

1. **Scatter:** However, the relationship is not perfectly linear. The data points are scattered around the trend, suggesting that other factors besides rainfall might influence yield.
2. **Saturation Point:** There might be a point where further increase in rainfall does not lead to a significant increase in yield. This could be due to factors like soil saturation or leaching of nutrients.

5.3 True Vs. Predicted Crop Yield

Figure 6 shows the comparison between true and predicted crop yields. The points on the plot represent the actual crop yields vs. the predicted values from the model. In the figure, it appears to be a strong positive correlation between true and predicted yields. As the true yield increases, the predicted yield also tends to increase. This is indicated by the general upward trend in the data points.

5.4 Feature Importance

The model also computes the importance of each feature (independent variable) in making predictions. This is useful for understanding which factors (e.g., temperature, rainfall, phosphorus) are most influential in predicting crop yield. In this case, the model predicts the crop type (label), but a more advanced version could predict actual yield by incorporating granular yield data. The Random Forest model learns the relationship between these features and the crop label, identifying patterns such as the ideal phosphorus and potassium levels, temperature range, and humidity for specific crops like rice. If actual yield data were available, the model could predict the crop yield (e.g., tons per hectare) directly rather than the crop type. This machine learning model aids in optimizing farming practices by understanding how environmental and soil conditions affect crop growth, offering valuable insights for

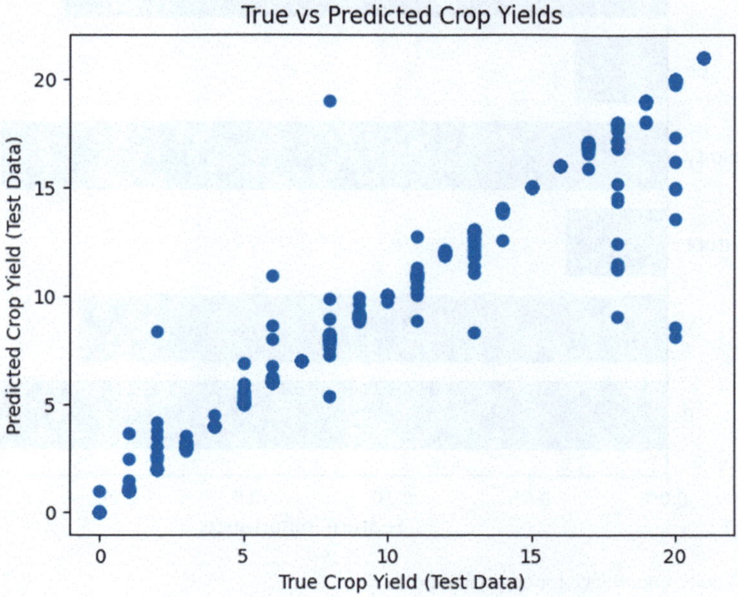

Fig. 6 True vs. predicted crop yield

precision agriculture. It helps farmers make informed decisions on crop management, irrigation, fertilization, and harvesting, ultimately improving productivity and sustainability. Figure 7 illustrates the feature importance for predicting crop yields. The length of each bar represents how much each feature (e.g., temperature, rainfall, phosphorus) contributes to the model's prediction of crop yield. Longer bars indicate more significant features, while shorter bars suggest less influence.

This section uses Random Forest Regression to predict crop yields based on environmental and soil nutrient variables such as phosphorus (P), potassium (K), temperature, humidity, pH, and rainfall. The model, trained on a dataset of these variables, helps predict the crop type (label), with potential adaptation for predicting actual yield if detailed yield data were available. By learning the relationship between these factors and the crop type, the model aids in precision agriculture, guiding farmers in decisions related to crop management, irrigation, fertilization, and harvesting. Ultimately, it optimizes farming practices and enhances productivity through data-driven insights.

The correlation heatmap in Fig. 8 provides insights into the relationships between various crop yield-related features. Nitrogen, phosphorus, potassium, and rainfall exhibit positive correlations with yield, suggesting that increasing their levels can contribute to higher crop yields. Temperature, on the other hand, shows a slight negative correlation, indicating that extreme temperatures can negatively impact

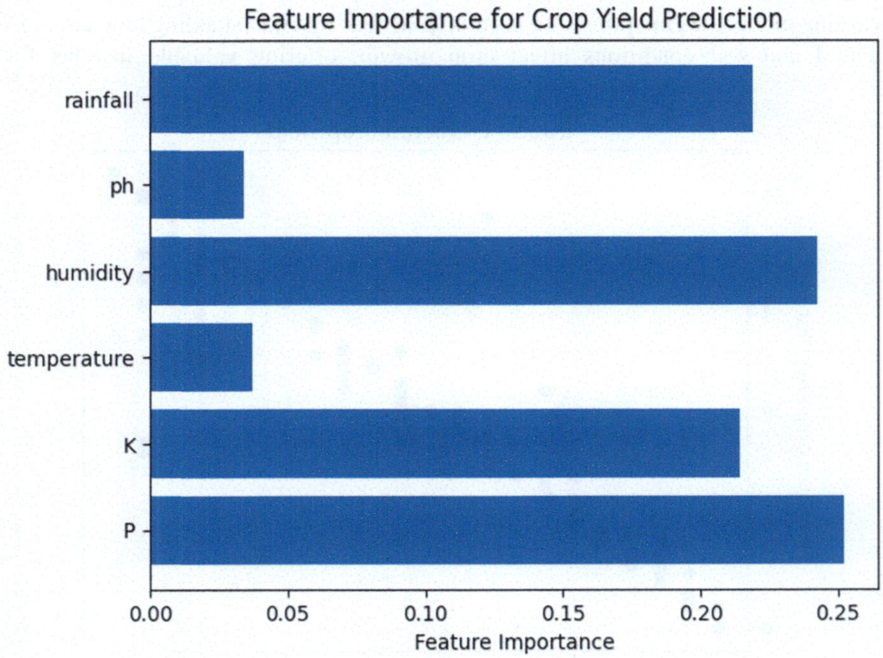

Fig. 7 Feature importance for crop yield prediction

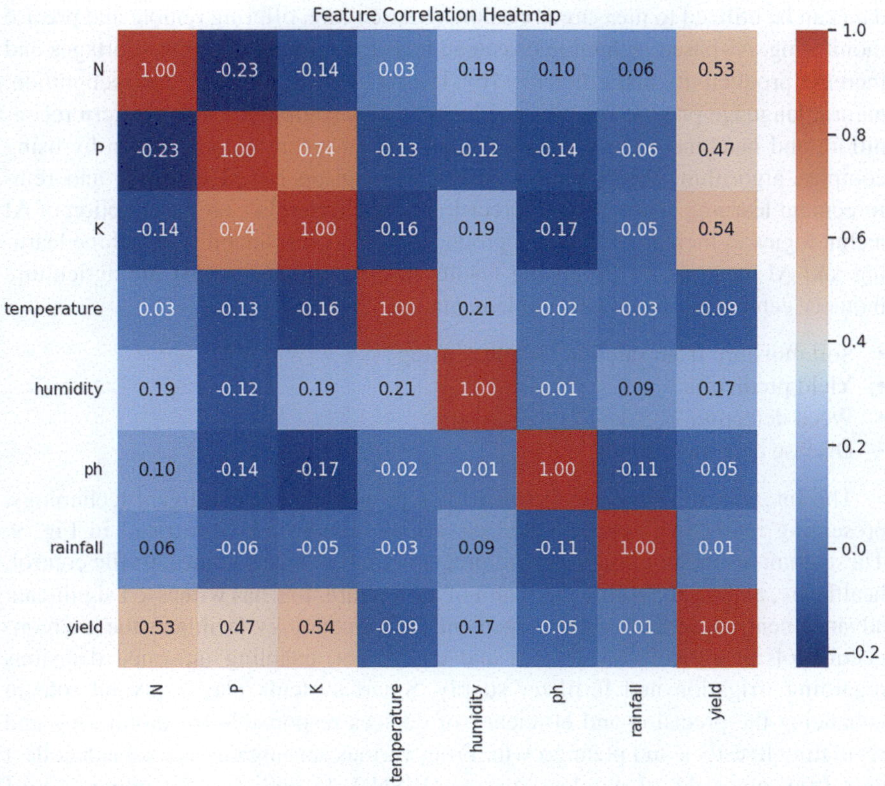

Fig. 8 Correlation heatmap

crop growth. The strong positive correlation between phosphorus and potassium highlights their synergistic effect on crop development.

6 Empowering Agriculture with AI: Innovative Smart Farming Practices for a Sustainable Future

The agriculture sector is increasingly using Artificial Intelligence (AI) to help farmers maximize yields, reduce input costs, and increase efficiency in their operations. This technology can monitor and detect problems in crops, optimize soil management, and automate crop harvesting. AI can learn from data and make decisions without explicit programming. Precision agriculture has utilized AI to optimize crop production, including controlling water, nutrients, and fertilizers [4]. With the growing population, the demand for agriculture increased, and AI has become an essential tool for examining patterns quickly and accurately, allowing businesses to utilize connected information. By implementing machine learning and AI, digital data can be utilized to measure precision in agriculture, offering remote and precise monitoring. AI-based technologies can address the problem of labor shortages and increase productivity and efficiency [6]. The AI cycle includes speech recognition, natural language processing, advanced mechanics, computer vision, pattern recognition, and machine learning. AI technologies can create a digital brain by using complex algorithms through directed learning, unsupervised learning, and reinforcement learning. The future of precision agriculture relies on the adoption of AI technologies to increase efficiency, productivity, and sustainability. Machine learning and AI have shown promising results in other countries, and the agriculture industry can benefit from their implementation [7].

- Soil moisture from satellite remote sensing
- Yield prediction
- Weed detection
- Disease detection

The Internet of Things (IoT) stands as a promising and intelligent technology, presenting innovative solutions across various domains, as illustrated in Fig. 9. These domains include smart agriculture, smart cities, smart homes, traffic control, healthcare, and more. Within the realm of agriculture, IoT has witnessed significant advancements in agricultural management. This technology facilitates the interconnection of all agricultural devices and equipment, enabling informed decisions regarding irrigation and fertilizer supply. Smart systems play a pivotal role in enhancing the precision and efficiency of devices responsible for monitoring and regulating livestock and plant growth. From various sensing devices we can collect data from networks of wireless sensors (WSNs). In addition, the integration of cloud services with IoT is critical to the analysis and processing of remote data,

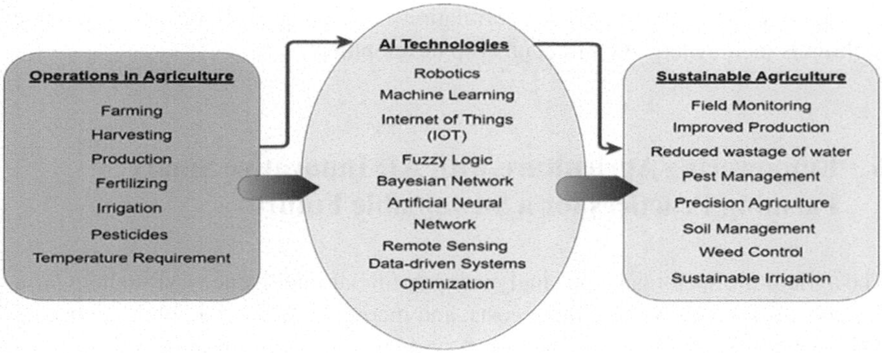

Fig. 9 AI in agriculture

Table 3 Application and implementation of modern farming techniques

S.no	Application	Description
1	Monitoring climate condition	Utilizes IoT sensors to monitor environmental factors such as humidity, rainfall, temperature, and soil moisture levels in real time
2	Crop and disease management	Involves the use of IoT technology to surveillance crop health and detect early signs of disease outbreaks or insect infestations. Data collected by drones and IoT devices is analyzed using machine learning techniques to provide farmers with warnings
3	Irrigation	IoT-enabled systems automate and improve irrigation processes by gathering data from meteorological sensors and soil moisture sensors
4	Fertilizer management	Soil sensors collect data on soil pH, nutrient concentrations, and moisture content, which is then used to provide fertilizer recommendations. IoT-enabled fertilizer injectors or spreaders apply precise nutrient dosages
5	Soil sampling	Automates soil sampling processes by using IoT-equipped equipment to collect data on soil texture, structure, moisture content, and nutrient levels. Information is wirelessly transmitted to a central database, where algorithms analyze the data and generate maps showing soil health and management suggestions

which in turn helps decision-making processes to carry out optimal solutions. Smart farm management requires the use of information and communication technology (ICT). The utilization of ground sensors, robots, autonomous automobiles, and other automated equipment are also necessary for effective smart farming. These smart systems require rapid internet access, position data, and satellites that provide essential images and contemporary mobile devices to operate successfully used real-time IoT. This involved covering large areas using images from the satellites and a lot of detectors, particularly in areas which were growing banana and paddy crops. This necessitated the installation of several sensors and satellite images in fields, particularly those that were cultivating paddy or banana crops. The system allowed for data analysis and decision-making, and it used a web server to communicate the findings to farmers.

In Table 3, the various applications and their implementation are discussed.

In Table 4, applications have been discussed with examples.

Flowchart of the work has been depicted in Fig. 10 and discussed as follows:

First data is collected from sensors, after collection it gets processed based on the requirement. Further recommendations and alerts are provided. Based on the recommendations farmers can make their decisions.

Table 4 Applications with examples

S.no	Application	Examples
1	Soil moisture monitoring	IoT sensors embedded in the soil to measure moisture levels at different depths Data is transmitted wirelessly to a central system for analysis Farmers receive real-time alerts and recommendations for irrigation scheduling based on soil moisture data
2	Crop health monitoring	Images of crops captured by drones equipped with multispectral cameras Images are analyzed by machine learning algorithms to look for indications of illnesses, pests, or dietary deficiencies Farmers receive actionable insights and recommendations for targeted interventions to maintain crop health
3	Precision irrigation	Automated irrigation systems adjust water application rates based on real-time data Precision watering minimizes water waste and ensures crops receive the right amount of moisture
4	Livestock tracking	IoT-enabled ear tags or collars monitor the location and behavior of livestock Data on grazing patterns, activity levels, and health status is collected and analyzed Farmers can optimize herd management practices and detect anomalies early for prompt intervention
5	Crop yield prediction	IoT sensors collect data on environmental conditions, soil health, and crop growth stages Predictive analytics models forecast crop yields based on historical data and current inputs Informed decisions can be made by farmers on harvesting schedules and resource allocation for maximum yield

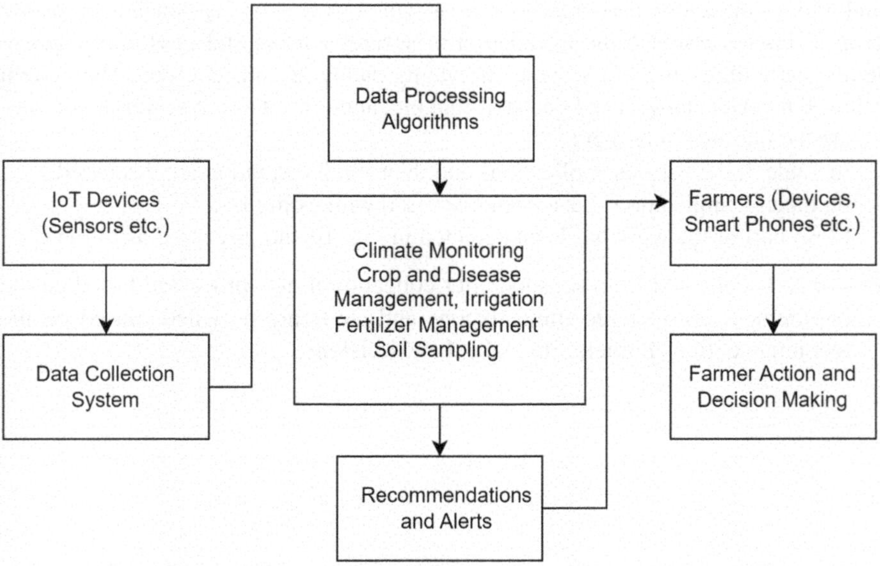

Fig. 10 Flowchart of transforming agriculture: maximizing farming efficiency through integrated IoT solutions

7 Smart Farming Applications

- *Advancing Artificial Intelligence:* Smart farming is moving away from large central servers and toward the use of smaller, autonomous devices like wireless sensors. This implies that decisions about farming may be made more quickly and effectively when needed.
- *Open-Source Solutions:* Rather than relying only on pricey proprietary systems, farmers will increasingly use open-source technologies. Everyone will be able to afford and have greater access to smart farming as a result. It will also facilitate information exchange and mutual aid among farmers.
- *Automated Farm Equipment:* In the future, most agricultural machinery will be like smart vehicles, driving themselves. They'll take care of everything from pest control to seed sowing using cutting-edge technology like computer vision. Thus, farmers will require technological know-how to compete with the competition.
- *Blockchain for Security:* Cyber attack defense will be necessary for smart farming, just like it is in other domains. Farmers' data will be safe and secure, thanks to blockchain technology, especially when it's shared online.
- *Data-Driven Agricultural Decisions:* Farmers will utilize data and machine learning to make data-driven agricultural decisions rather than supposing to make educated guesses. This entails examining long-term climatic trends in order to make more informed choices that will increase agricultural yields.
- *5G for Internet Access:* Smart farming in the future will be significantly dependent on 5G wireless technologies. Faster and more dependable connections will be made possible by it, which is crucial for managing intelligent equipment like tractors and drones as well as monitoring crops in vast fields [8].

8 Conclusion and Future Directions

In rural areas, agricultural families are grappling with the loss of the succeeding generation of farmers. This phenomenon is attributed to challenges such as escalating agricultural costs, inadequate per capita production, deficient soil maintenance, and a trend toward alternative occupations that either do not involve farming or offer more lucrative prospects than traditional agriculture. The lack of appropriate infrastructure makes using this technology quite difficult. Therefore, we should focus on the constraints and issues with data processing. We need to carefully link ground sensors to computers in an economical method to get the desired output. This will help make precision farming work well, even in places that might face challenges. Machine learning, prediction algorithms, and strong data analytics capabilities are necessary to extract valuable insights from gathered data. IoT systems in agriculture face a significant maintenance problem when it comes to hardware because of the frequent use and probable rapid deterioration of sensors. It is imperative to develop intelligent agricultural applications designed for field use. Business owners or farm managers should have the capability to access data on-site or remotely through smartphones or desktop computers. The effective operation of smart farming applications relies on a well-established internal infrastructure. An in-depth understanding of security issues and data privacy within smart farming is essential.

References

1. Unal, Z. (2020). Smart farming becomes even smarter with deep learning—a bibliographical analysis. *IEEE Access, 8*, 105587–105609. https://doi.org/10.1109/ACCESS.2020.3000175
2. Ghosh, S., & Kavitha, R. (2024). AI and ML in Agriculture. International Journal of Research Publication and Reviews, 5(3), 310–316
3. Joseph, R. B., Lakshmi, M. B., Suresh, S., & Sunder, R. (2020). Innovative analysis of precision farming techniques with artificial intelligence. *2nd International Conference on Innovative Mechanisms for Industry Applications (ICIMIA 2020):Conference Proceedings*, Icimia, pp. 353–358. https://doi.org/10.1109/ICIMIA48430.2020.9074937.
4. Jha, K., et al. (2019). A comprehensive review on automation in agriculture using artificial intelligence. *Artificial Intelligence in Agriculture, 2*, 1–12.
5. Kumar, M., Bajaj, K., Sharma, B., & Narang, S. (2022). A comparative performance assessment of optimized multilevel ensemble learning model with existing classifier models. *Big Data, 10*(5), 371–387.
6. Bajaj, K., Jain, S., & Singh, R. (2023). Context-aware offloading for IoT application using fog-cloud computing. *International Journal of Electrical and Electronics Research, 11*(1), 69–83.
7. Smith, J., Petrovic, P., Rose, M., De Souz, C., Muller, L., Nowak, B., & Martinez, J. (2021). Placeholder text: A study. *The Journal of Citation Styles, 3*. https://doi.org/10.11591/ijpeds.v14.i2.pp1044-1054
8. Qazi, S., Khawaja, B. A., & Farooq, Q. U. (2022). IoT-equipped and AI-enabled next generation smart agriculture: A critical review, current challenges and future trends. *IEEE Access, 10*, 21219–21235. https://doi.org/10.1109/ACCESS.2022.3152544
9. Idrees, S. M., & Nowostawski, M. (Eds.). (2024). *Blockchain transformations: Navigating the decentralized protocols era*. Springer.
10. Singh, I., Kaur, A., Agarwal, P., & Idrees, S. M. (2024). Enhancing security and transparency in online voting through blockchain decentralization. *SN Computer Science, 5*(7), 921.
11. Satybaldy, A., Subedi, A., & Idrees, S. M. (2024). Decentralized key management for digital identity wallets. In *Blockchain transformations: Navigating the decentralized protocols era* (pp. 47–58). Cham.

Enhanced Blockchain Architecture for Wheat Crop Supply Chain Management

K. Geetha, N. Sasikaladevi, and S. Aarthi

1 Introduction

The wheat crop supply chain encounters numerous problems [1] that makes blockchain technology as convincing solution. The global food distribution network is very complicated, with problems with trust, transparency, and being able to track food. This makes blockchain integration all the more appealing. Traceability is the most critical requirement for blockchain implementation in the wheat crop supply chain [3–5]. Besides traceability, transparency, and operational efficiency [2–5], Blockchain offers improved security measures. This resilience in tamper proof and cyber intrusions is an inherent feature of a decentralized architecture of blockchain [2]. Because of the nature of cryptography-based block inter-linkages between consecutive blocks in the chain, alteration in data will make the entire data repository more challenging to modify surreptitiously.

1.1 Protocols of Consensus

Consensus protocols are needed for blockchain networks to work because they let everyone agree on whether a transaction is valid and keep a shared record of its status. This gets rid of the need for an administrative unit. The number of consensus protocols that have emerged has been multiple, each having its unique method of allowing mutual understanding among network consumers.

Proof-of-Work (PoW): Bitcoin's innovative consensus protocol requires miners to compete to solve computationally difficult mathematical puzzles. The successful

K. Geetha · N. Sasikaladevi · S. Aarthi (✉)
SASTRA Deemed University, Thanjavur, Tamil Nadu, India

S. M. Idrees et al. (eds.), *Blockchain and Machine Learning Innovations*, Signals and Communication Technology,
https://doi.org/10.1007/978-3-031-88095-7_5

miner may add a new blockchain block as well as receive cryptocurrency as a reward. PoW is energy-intensive and can lead to centralization due to the large computational power of some miners, but it also ensures network safety by making network attacks computationally expensive.

Proof-of-Stake (PoS): The selection of validators for the creation of new blocks depends on the amount of cryptocurrency they possess and are willing to "stake" as a deposit. This protocol serves as an alternative to the PoW consensus algorithm. PoW is computationally intensive when compared with the PoS algorithm; hence, it processes transactions faster. However, the PoW algorithm may lead to centralization, as it relies on the probability of selecting significant amounts of funds from individual users to validate each transaction.

1.2 Related Works

Several researchers looked into how to combine the point-of-sale (PoS) and distributed ledger approaches in the management of agriculture's supply chains. They were especially interested in how to lower operational costs, make transactions more transparent, and make sure that all transactions could be tracked. Farooq et al. listed the scope of distributed ledger technology while supervising the supply chain in the transactions of crop management. The corresponding article investigates the computational efficiency of blockchain in crop supply chains, highlighting the trust and transparency of decentralized and immutable ledger schemes [6]. Madhumitha et al. discussed the challenges of utilizing blockchain in crop management, including the implementation challenges and lower transaction costs [7]. Zhang et al. analyzed various works of literature on blockchain technology that focus on data sharing and supply chain management by exploring different consensus algorithms such as PoS in terms of security and privacy [8]. Similarly, Duan et al. proposed a crop supply chain integrated with PoS consensus, enhancing transparency and provenance, thus avoiding fraudulent transactions and ensuring crop safety [9]. Galvis et al. analyzed several papers systematically concerning the existing schemes in terms of traceable solutions in agri-food supply chains [10, 11]. The results impact the comprehensive review of crop product authentication, visibility of crop supply chains, and quality control with the distinct architectures of blockchain technology. Collectively, these above research works highlight the requirement of blockchain and PoS mechanisms in the modernization of crop supply chains and offer insight into the possible benefits, challenges, and opportunities surrounding their implementation in agricultural ecosystems.

1.3 Motivation

It is suggested that a blockchain-based wheat crop supply chain management system be put into use in real-time to help with problems like traceability, transparency, and efficiency that already exist in crop supply chains and inventory management. Blockchain technology allows for decentralization and an immutable record. Hence, the proposed scheme enables the transparent and safe tracking of wheat from farm to consumer.

Integrating blockchain technology can help maintain transaction and certification integrity, accelerate transactions, and increase trust among stakeholders. This will form a sustainable and resilient environment for the wheat crop supply chain, resulting in higher accountability and value creation for all participants. The following Table 1 lists the differences between the consensus algorithms of blockchain technology in terms of energy efficiency, scalability, security, and privacy. Fig. 1 above depicts the supply chain management in the crop processing model.

1.4 Significant Contribution

The following limitations in the existing wheat supply chain management system should be addressed. Transparency is a significant issue, as tracing the origin of wheat is difficult, which could result in deception. Furthermore, security inefficiencies and functional disabilities occur due to manual observation and paper-based communication. It drastically increases the communication costs and overhead in each transaction due to the involvement of middlemen, generating the impedance of transparency in wheat crop supply chain management. Subsequently, improper

Table 1 Difference between PoW and PoS

Criteria	Proof of work (PoW)	Proof of stake (PoS)
Consensus mechanism	Relies on computational competition among miners to validate blocks	Selects validators based on their stake in the network to validate blocks
Energy efficiency	Criticized for high energy consumption due to computational puzzles	Generally consumes less energy compared to PoW because of the absence of complex computations
Security and reliability	Highly secure but faces reliability challenges	Offers security based on validators' stake, with improved reliability

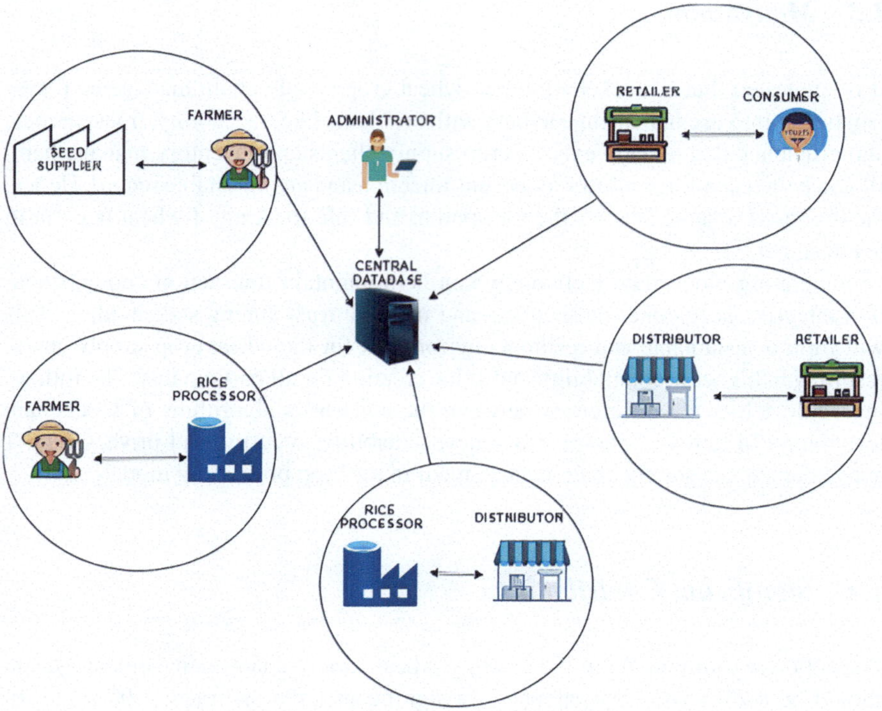

Fig. 1 Supply chain management

regulatory compliance restricts the consistency and reliability of each entity involved in the crop supply chains. Collectively, the above limitations reduce the trust and efficacy of the wheat supply chains with increasing operational costs. Therefore, providing a scheme that eliminates middlemen and investigates equitable distribution is important. The fundamental research objectives comprise the following:

1. ***Elimination of middlemen:*** To optimize the distribution of crops by eliminating several middlemen that reduce the expenses and affirm more equitable prices for both the producers and consumers [12]. This is based on utilizing the direct crop transactions between the farmers and retailers to decrease the requirement of intermittent markups and delays.
2. ***Equitable Distribution:*** All stakeholders must gain the provision of fair pricing and accessibility through equitable distribution. The output of the distributed ledger technology and the smart contract [13] will introduce the deployment of fair policies across all stakeholders that encourage participants, from small-scale crop producers to marginal communities. The suggested plan also adds more types of distributed channels to areas that aren't well served, which improves crop security and productivity [14].
3. ***Efficiency Improvement:*** Blockchain with a PoS consensus algorithm introduces transparency, tracking inventory, and optimizing routes in wheat supply

chain management [15]. The automation process through smart contracts decreases the operational costs and delays due to the middlemen [16]. Moreover, the proposed scheme allows real-time decision-making with minimal resource utilization.

4. ***Transparency Promotion:*** Promoting trust and the integrity of data in the wheat supply chain is essential in a critical and untrusted environment. With the PoS consensus algorithm and blockchain technology, transactions can't be changed, so attackers can't change the data in the crop supply chain. This makes sure that every transaction, from the farm to the store, is checked. Ethical concerns and sustainability in crop production and distribution are also kept an eye on by digital signatures from everyone involved, certification standards set by reputable organizations, and clear pricing systems that let customers know about decisions being made in real-time in crop supply chains.

The organization of the manuscript is arranged as described below:

Secttion "Experimental Research" mentions the research objective for developing a blockchain-based solution to manage the crop supply chain, particularly wheat crops. Section "Results and Discussions" illustrates the experimental design and methodology involved in the proposed scheme. Section "Conclusion and Future Work" describes the analysis of the proposed scheme against the current state of the art. The last section enumerates the conclusion and future improvements. In the appendix, we have included the code for the PoS consensus algorithm for wheat crop supply chains.

2 Experimental Research

The proposed system comprises two distinct modules, including Module 1, which is responsible for inputting regulated wheat prices, and Module 2, which manages verification and transactions. Furthermore, smart contracts are essential for the maintenance of system reliability and the implementation of security measures. They enhance the system's overall integrity and credibility by enforcing predefined rules and facilitating secure transactions.

2.1 Smart Contract Development

Smart contracts are digital agreements between producers and retailers that help manage the supply chain using the blockchain. The Ethereum platform automates these agreements and makes sure that transactions are honest. Therefore, these specific needs from the supply chain would require the verification process for authentic products, the processing of a payment, and then the transfer of wheat. The smart contract verifies the conditions specified for quantity, quality, and price of crops at the time of transaction initiation. The contract then autonomously carries out the transaction if these criteria are met, thus ensuring that payment is made using cryptocurrency and transferring ownership of wheat from the farmer to the retailer. The technology allows cutting out intermediaries from the process, facilitating and making it cheaper when done by traditional supply chains. As a result, an immutable and transparent blockchain approach ensures transaction process authenticity and integrity reducing fraud and dispute occurrences. Through Ethereum, with a decentralized network of its intelligent chain, a user ensures clear just trade practices from their point to the retailer where all transactions are followed or passed down by the wheat flowing down. Finally, this blockchain-driven solution will make the wheat supply chain more efficient, transparent, and reliable for farmers, retailers, and consumers. Figure 2 depicts the code snippet provided for wheat supply chains.

Furthermore, Fig. 3 illustrates the overall implementation of a smart contract between two parties including the creation of blocks, transactions between two parties, and adding the corresponding block in the blockchain.

```
//SPDX-License-Identifier: MIT
pragma solidity >=0.7.0 <0.9.0;
//SPDX-License-Identifier: MIT
contract StructMapping {
        mapping(address => uint)
        public balanceReceived;
function getBalance() public view returns(uint) {
        return address(this).balance;
}
function sendMoney() public payable {
        balanceReceived[msg.sender] += msg.value;
}
function withdrawAllMoney(address payable _to) public {
        uint balanceToSend = balanceReceived[msg.sender];
        balanceReceived[msg.sender] = 0;
        _to.transfer(balanceToSend);
}
receive() external payable {
        sendMoney();
}
}
```

Fig. 2 Code snippet for smart contract

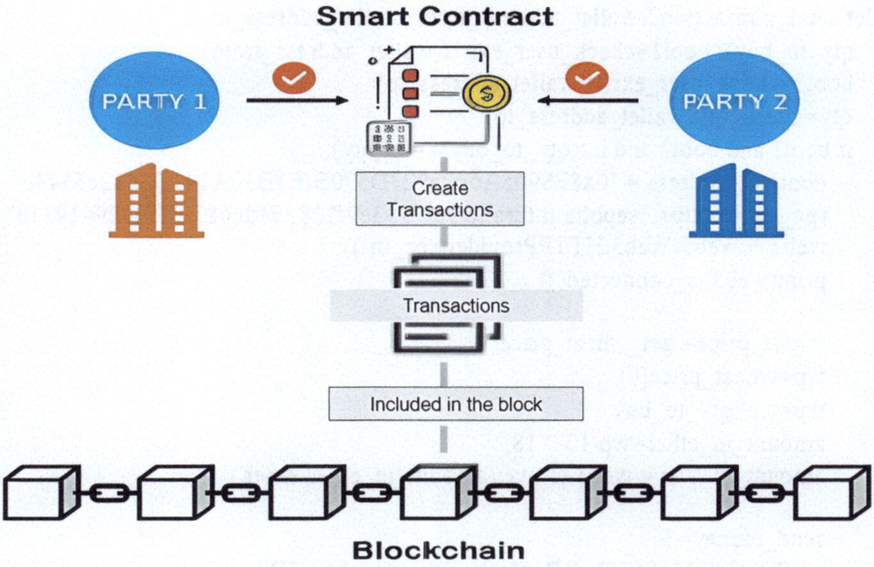

Fig. 3 Flowchart of smart contract implementation

2.2 *User Interface Design*

In the complex wheat supply chain management, developing a user interface (UI) specifically for farmers and retailers would be a significant consideration. This is due to the critical role these UIs play in facilitating smooth interaction between stakeholders and blockchain systems via web-based platforms. The proposed scheme includes the front-line User Interface (UI) that enables the preferences of farmers and consumers to fulfill their transactions effectively. For instance, the UI on the side of the farmer should contain more features for finding the retailers to sell the crop without any middlemen. It must ensure smooth navigation of the transaction between the farmer and retailer. Below is a detailed description of the user interface. First, it should show crop information concerning its quality, source, and volume. Secondly, functions such as tracking orders and managing inventories should be permitted to enhance the effectiveness of operations and the optimization of resources. These functions assist the farmers in interfacing their inventories with their order progress. Figure 4 lists the code snippet for the transaction that sends the money from the wallet.

After that, there should be an easy-to-use interface that lets retailers focus on the tasks that allow them to see the displayed crop categories, specific crop information, and product details and make safe purchases. With that, we have given the authentication mechanisms to verify the participants' identity along with the crop purchase information, and therefore events such as adulteration and the purchase of fake goods are minimized. These interfaces allow for fire-blocking accessibility and effectiveness to be built on as the design process evolves while the security and privacy of the crop supply chains are well preserved.

```
def send_transaction2(wallet_address_from, wallet_address_to,
  qty_to_buy): bool1=check_user_exists(wallet_address_from)
  bool2=check_user_exists(wallet_address_to)
  qty=check_qty(wallet_address_to)
  if bool1 and bool2 and int(qty_to_buy)>=int(qty):
    contract_address = "0x8F8905A6F7a927D570BfE1B30A147D3482c6344e"
    rpc_url = "https://sepolia.infura.io/v3/7933e9f128364db6870f7f902941931d"
    web3 = Web3(Web3.HTTPProvider(rpc_url))
    print(web3.is_connected())

    wheat_price= get_wheat_price()
    wp=wheat_price[0]
    wp=wp*qty_to_buy
    amount_in_ether=wp/10**18;
    amount_in_wei = web3.to_wei(amount_in_ether,'ether')

    send_money={
      'chainId': 11155111, # Replace with your chain ID
      'value': amount_in_wei,
      'gas': 250000, # Adjust gas limit as needed
      'gasPrice': web3.to_wei('50', 'gwei'), # Adjust gas price as needed
    'nonce': web3.eth.get_transaction_count(wallet_address_from), }
    withdraw_all_money={
      'chainId': 11155111, # Replace with your chain ID
      'gas': 250000, # Adjust gas limit as needed
      'gasPrice': web3.to_wei('50', 'gwei'), # Adjust gas price as needed
    'nonce': web3.eth.get_transaction_count(wallet_address_from), }
```

Fig. 4 Code snippet for transaction

2.3 Integration with Off-Chain Services for Wheat Prices

Improving the functionality and dependability of smart contracts in the context of managing the wheat supply chain means incorporating web services that get real-world data like the current price of wheat. This process involves utilizing web scraping techniques to extract relevant data from various online sources and feed it into smart contracts to inform transaction decisions. Web scraping of crop prices entails hacking agricultural market sites for relevant price data through the use of libraries such as Beautiful Soup from Python. Such resources allow stakeholders to improve their decision-making process by obtaining the best and most up-to-date prices for the products. There is also the possibility of web scraping for price changes; hence, the stakeholders will know the right time to change their plans [17]. Afterward, the blockchain network incorporates this data into smart contracts. Smart contracts are self-executing the rules written in them. They operate autonomously and always respond to a transaction; for instance, they can use various methods when certain conditions are satisfied [18].

In a smart contract, a specific key is meant for a designated party, and it is used to digitally sign or stamp the contract. We refer to this process as "digital signatures," a term that partially stems from the mathematical constraints inherent in standard signatures. Simply put, through the help of certain mathematical algorithms, each user can come up with a signature unique to himself, which depends on his private key. The transaction body then incorporates these signatures along with the corresponding data signature. Figure 5 illustrates the complete workflow of the wheat supply chain.

Smart contracts use these digital signatures to confirm that the participants are trustworthy and that whoever is permitted to perform the transactions does so safely over the blockchain network. As such, this means that transactions can only be carried out by appropriate parties, adding to the reliability and safety of the entire ecosystem of smart contracts. Similarly, we can provision other types of smart contracts to validate data, ensuring consistency and accuracy, particularly when acquiring information from third parties. To do this, we might use consensus cross-validation of sources, gather information from three different sources, and even check and react to the data's current state regularly.

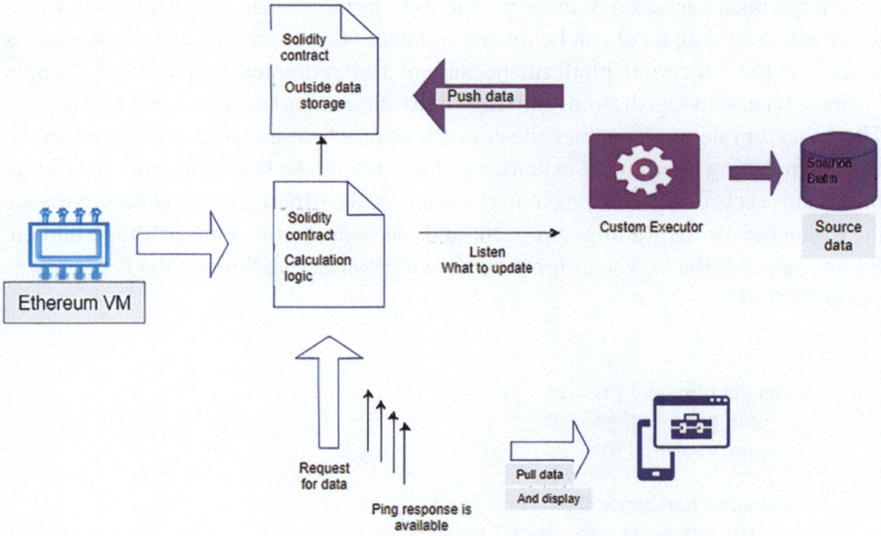

Fig. 5 Flowchart of wheat price feeding

2.4 Deployment on Ethereum Network

The proposed scheme is to use the Ethereum blockchain to deploy smart contracts and build DApps, as it has enough security and the ability to scale. The proposed scheme has utilized the tracking and inventory management of the wheat supply chains from the producer to the consumer. We constructed it using self-executing agreements with predetermined and encoded conditions and terms, also known as smart contracts. When the terms and conditions meet automatically, these smart contracts execute. This specific feature of the system eliminates the use of middlemen and enables the stakeholders to engage in trustless transactions.

Additionally, smart contracts offer a way to log wheat supply chains so that DLT cannot alter the records. ECC encrypts every transaction and assures data integrity using digital signature algorithms. Moreover, the onward ledger makes it possible to develop trusted relationships and enables people to track the crops they bought, as it allows viewers to follow their crops through the transactions to the final buyers. Through the mechanism of proof-of-stake, we have improved the parameters, such as gas costs and transaction parameters. The gas expenses are mandatory for miners who validate the transactions and process them on the Ethereum network. When gas costs and transaction parameters are set as per the consensus algorithm, they can guarantee that such transactions will be carried out in the shortest time possible and cheaper. Figure 6 includes the code snippet for the gas cost calculation for each transaction in the Ethereum virtual machine.

For instance, smart contracts can be developed to update wheat inventory levels once it has been harvested or make payments to farmers when they deliver the wheat to the retailer. Also, there can be tokens created to represent physical assets such as wheat on the Ethereum platform because it has promoted tokenization. Supply chains, ownership registration, and transfer to another party can utilize these tokens. The Ethereum network enables the conversion of wheat assets into a token format, thereby reducing operational expenses and shortening the trading period. By changing wheat assets into tokens, stakeholders can use the Ethereum blockchain, enhance the efficiency of the trading approach, and cut back on the administrative burden. Figure 7 depicts the flowchart for validating the transaction through the PoS consensus algorithm.

```
# Get gas price and gas used
    gas_price = tx['gasPrice']
    gas_used = tx['gas']

# Calculate transaction fee
    transaction_fee = gas_price * gas_used

    print("Gas Fee:", web3.from_wei(gas_used * gas_price, 'ether'), "ETH")
    print("Transaction Fee:", web3.from_wei(transaction_fee, 'ether'), "ETH")
```

Fig. 6 Code snippet for the gas costs calculation

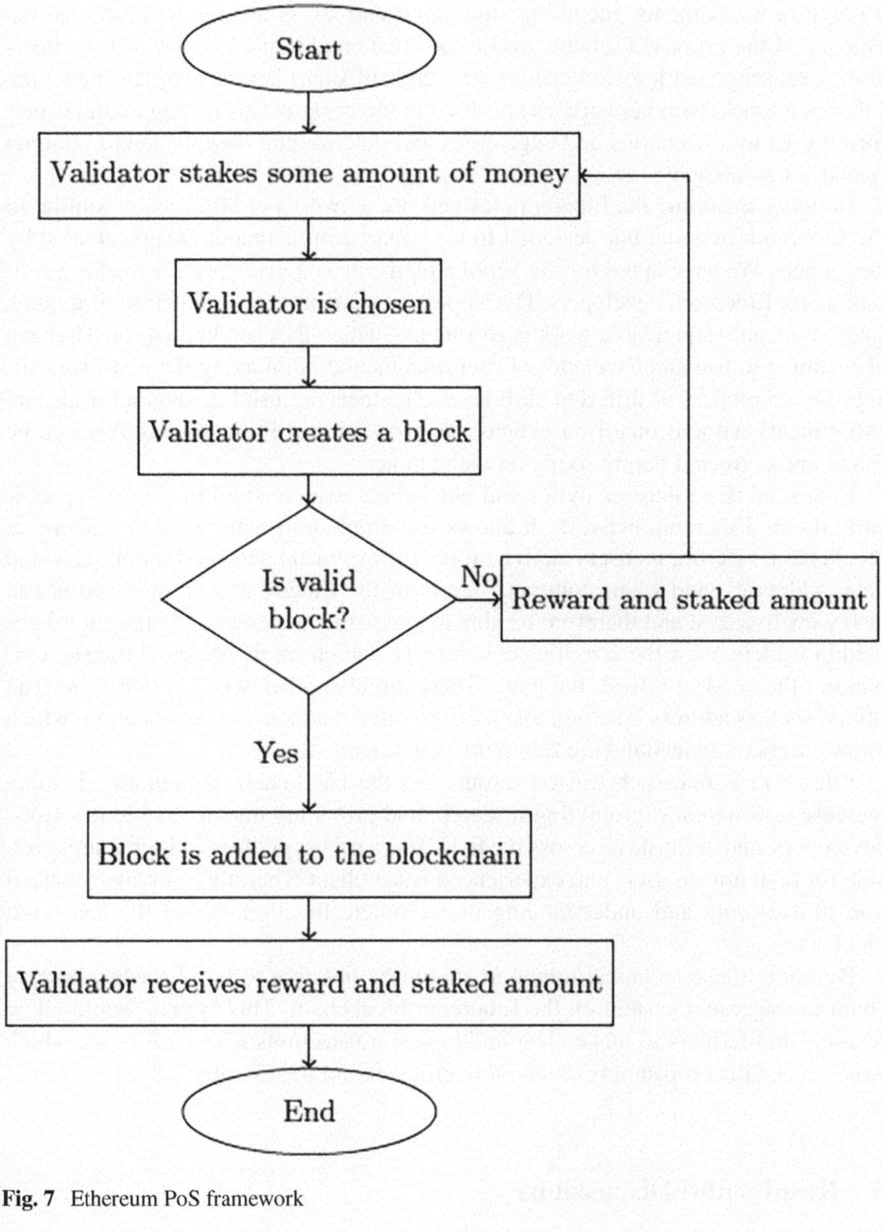

Fig. 7 Ethereum PoS framework

2.5 *Testing and Deployment*

Testing the entire wheat supply chain management system, including smart contracts, user interfaces, and off-chain components, is crucial to ensure its functionality, stability, and potential for expansion. This step is quite detailed in that it requires a range of assessments, including simulation and stress testing, to guarantee the efficacy of the proposed scheme under specified conditions. We also test the operability, security, and logical business structures of smart contracts operating on the Ethereum blockchain network, as specified in the design. This testing includes performing various scenarios and edge cases and determining how the smart contract would act as designed.

Roughly speaking, the Ethereum testnets are networks of blockchains similar to the Ethereum mainnet but designed to test smart contracts and DApps created by developers. We have opted for the Sepolia platform as it gives perfect working conditions for Ethereum developers. Developers consistently rely on their high-quality, stress-free, and dependable work performance, which they are known for. They are also simple to use and have a lot of user documentation, making the platform suitable for developers of different skill levels. Testnets are used to conduct trials and experiments without incurring expense and loss of real Ether so mistakes can be found and corrected before fixing on the mainnet.

Etherscan is a robust analytics and blockchain explorer tool that works specifically on the Ethereum network. It allows for simple transactions on the Ethereum blockchain, offering its users analytic tools for exploring and verifying transaction data, addresses, and smart contracts. Users of the Etherscan web application can also view live data and therefore be able to check the movement of Ether or tokens held in wallets, view the activities of contracts, and check the status of transactions without the need to refresh the page. There are also other services that Etherscan offers, such as address labeling, token information, and contract verification, which allows users to understand the Ethereum ecosystem.

Etherscan is one such trusted resource in the blockchain community, helping increase transparency, promoting research, and providing data to enable investors, developers, and enthusiasts across the field. It is a robust platform with features suitable for both novice users and experienced researchers. Therefore, it plays a crucial role in exploring and understanding the complete functionality of the Ethereum blockchain.

By doing these technical things, it should be possible to build a wheat supply chain management system on the Ethereum blockchain. This system would allow retailers and farmers to make clear and honest transactions with each other, which would make the crop supply chain more efficient and trustworthy.

3 Results and Discussions

The blockchain-based proposed scheme has solved the efficiency problem of transactions across various domains, including the wheat crop supply chain. Most transactions between farmers and retailers were inefficient, complex, and timely; they involved many parties before the incorporation of any blockchain-based system. People-oriented blockchains, like MetaMask, have completely changed the way transactions are processed in terms of speed, cost, and efficiency. The proposed plan uses these blockchains. The supply process of wheat crops reveals another hurdle in the form of transaction delays since they would often take several days before completion in the pre-blockchain era. Multiple middlemen compounded this layer of complexity, leading to unnecessary documentation, excessive checking, and slow payment completion. These activities interrupted the efficient distribution of the goods; however, they resulted in unnecessary expenses to all the parties. The weak points were the lack of a transparent and immutable ledger system, which resulted in trust issues and misunderstandings leading to further costs and time in resolving the disputes. Figure 8 visualizes the number of transactions in crop supply chains within 12 s.

PoS-based blockchains headed by Ethereum impart several features in transaction processing. PoS is different from other types of crypto because it uses different security methods. Unlike traditional PoW, where computers use their resources to fight, PoS consensus uses validators who stake their coins on the network to make sure transactions are real. This methodology decreases mining-related energy use but improves scalability and transactions per second. With the recently concluded Ethereum 2.0 upgrade aimed at changing the platform's consensus mechanism to one that is proof-of-stake, it's expected that there will be more efficiency in transactions that are executed. Another contributor to this efficiency revolution is MetaMask, one of the most popular Ethereum-based cryptocurrency wallets. MetaMask simplifies the process for users. MetaMask simplifies everything by enabling users to manage their digital assets and conduct straightforward transactions from a single interface. Because of its simplicity, MetaMask allows users to engage more actively in the expanding blockchain economy, for example, in the wheat supply chain, with a handful of mouse clicks.

Because of the integration of blockchain technology, the wheat crop supply chain transactions have experienced a transformation in their durations. The time that elapses from the start of a transaction to its completion has reduced from several

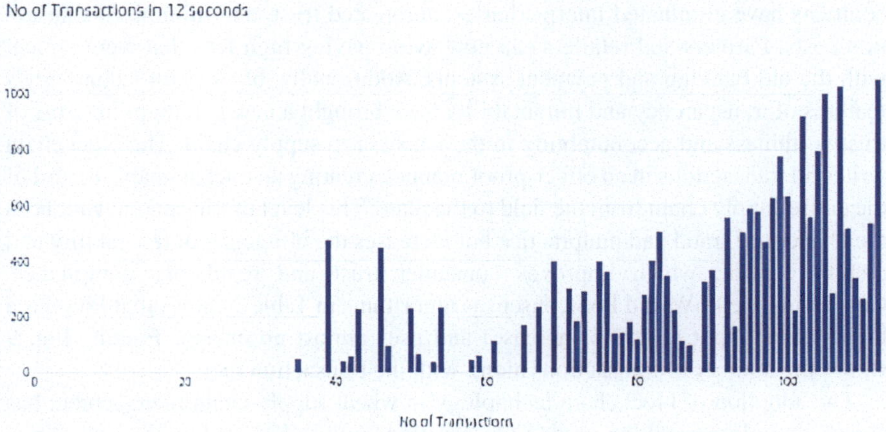

Fig. 8 Number of transactions in 12 s

Table 2 Comparison of throughput and energy consumption in various consensus protocols

Consensus protocol	Throughput (transactions per second)	Energy consumption
PoW	7 tps	1 billion
PoS	25 tps	1000

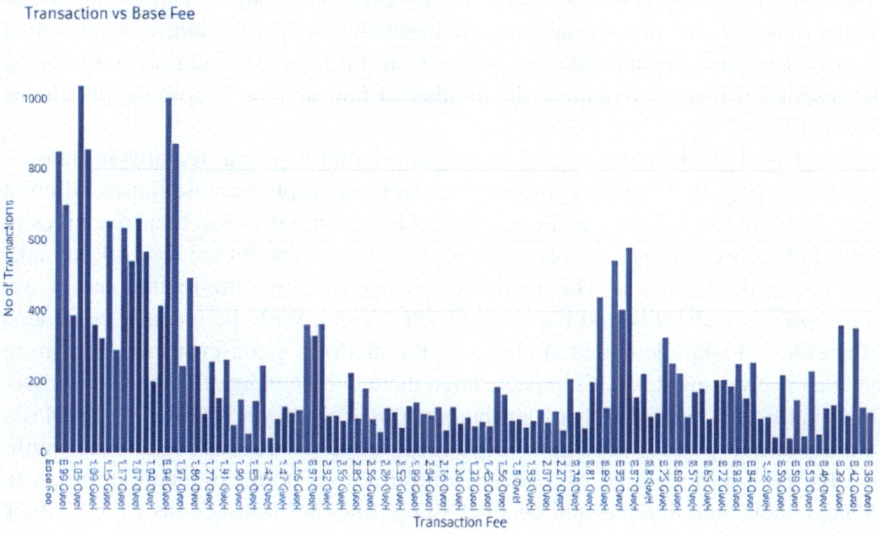

Fig. 9 Transaction vs. base fee

hours to a few seconds. The speed of transactions has therefore fastened the flow of the commodities in the supply chain and has enabled the producers and traders to quickly adjust to the changes in the market and its demands. In addition, participants in the wheat supply chain have also benefitted from the relatively low operational costs as a result of the deployment of the PoS consensus mechanisms. Smart contracts have eliminated intermediaries, automated trust, and minimized transaction costs. Farmers and retailers can now avoid paying high fees that were typical with the old banking and payment systems. Additionally, blockchain technology's features of transparency and immutability have brought a new paradigm in terms of trustworthiness and accountability in the wheat crop supply chain. The blockchain writes all transactions in an effect-proof manner, creating an unchangeable record of the entire supply chain from the field to the plate. This level of transparency reduces the chances of fraud and malpractice but increases the efficiency of traceability and quality control, which improves consumer trust and regulatory compliance. Comparing the PoW and PoS consensus algorithms in Table 2 shows that PoS has a higher throughput (a 257% increase) and uses almost no energy. Finally, Fig. 9 shows the number of transactions along with the transaction fees.

The adoption of blockchain technology in wheat supply chain management has significantly improved the speed of transactions to just 10 s and, at the same time, reduced the cost of transactions by 20%. This improvement makes for tremendous growth in efficiency and cost-beneficial effects on the supply chain, which makes for enhanced processes and better economic performance.

3.1 *Practical Scenario*

A problem must arise for wheat farmers in a rural farming village whereby they struggle to sell their produce to fair markets and equally delay payments. Earlier, they had to sell through agents who cheated them on the prices of produce and paid them very late. Further, retailers and consumers based in the urban areas were detached from the brands, due to quality and authenticity issues, which in turn raised the cost occasioned by poor supply chain operation. With the implementation of the proposed blockchain-based supply chain system, these challenges are addressed effectively:

- **Farmers:** Using the above-described blockchain platform, farmers provide only actual information about yields the ways they have been produced, and certifications. Payments are made directly to the farmers based on the agreed prices that smart contracts release immediately after a retailer acknowledges receipt of the shipment. Another advantage is that the platform assists farmers in knowing the market trends of different crops, hence helping the farmers plan the subsequent crops better.
- **Retailers:** Enhanced traceability is good for retailers: the blockchain provides data about every step from the farm to the store where the rice is sold. They answer the question of whether the rice they are offering is organic or of quality standard, which the consumers are willing to pay a premium price for such qualities.
- **Consumers:** Mainly, consumers use a QR code at the point of purchase to enquire about the product on the packaging material. This makes it possible for the buyer to know the source of the rice and the process of producing it, plus the transportation history making confidence in the rice.
- **Supply Chain Efficiency:** A flood affects transportation in one area, while the proposed blockchain solution finds other well-located wheat batches immediately. To avoid the risk of product losses, retailers redirect their orders from their original allocated places, thus making prices somewhat stable to the end consumer.

The proposed system, utilizing the Proof-of-Stake consensus algorithm, can decrease transaction processing time and lower transaction costs by 20%. The proposed supply chain management is appropriate for optimizing logistics and inventory management in crop processing, owing to enhanced throughput and reduced transaction costs. The projected gas for the suggested approach is likewise inferior to that of the PoW consensus process.

4 Conclusion and Future Work

There are quite a few areas where upgrades have not been fully explored, including how the wheat supply chain can be managed with blockchain technology powered by Ethereum's Proof-of-Stake (PoS) consensus mechanism. Potential upgrades aim to achieve seamless connectivity across different blockchain networks; greater development in this technology might come from enhancing data exchange and interactions between various participants in the wheat supply chain. The applicability of smart contracts to business operations like payment processing, quality assurance, and compliance monitoring has also improved. Data gathering and analysis could also benefit from the Ethereum blockchain, revealing insights into better supply chain management, consumer needs, and industry developments that are necessary for a data-driven decision-making process powered by AI models.

However, with these improvements, there are also noticeable challenges. Regulatory compliance emerges as a major concern for industries under direct scrutiny by regulatory bodies. Blockchain technologies present a significant trade-off between safeguarding confidential business information and adhering to regulations. Implementing robust privacy-preserving measures and encryption mechanisms is crucial to controlling access to sensitive information. To get the most out of using blockchain technology in the wheat supply chain, stakeholders need to think about several factors, such as network security and integrity, the complexity of the crop supply chain, and cost and resource limitations. Those who want to use blockchain technology for agrarian reforms can deal with these problems by planning strategically and coming up with new ideas together.

References

1. Amentae, T. K., Hamo, T. K., Gebresenbet, G., & Ljungberg, D. (2017). Exploring wheat value chain focusing on market performance, post-harvest loss, and supply chain management in Ethiopia: The case of Arsi to Finfinnee market chain. *Journal of Agricultural Science, 9*(8), 22.
2. Idrees, S. M., & Nowostawski, M. (Eds.). (2024). *Blockchain transformations: Navigating the decentralized protocols era*. Springer.
3. Idrees, S. M., Nowostawski, M., Jameel, R., & Mourya, A. K. (2021). Privacy-preserving infrastructure for health information systems. In *Data protection and privacy in healthcare* (pp. 109–129). CRC Press.
4. Satybaldy, A., Subedi, A., & Idrees, S. M. (2024). Decentralized key management for digital identity wallets. In *Blockchain transformations: Navigating the decentralized protocols era* (pp. 47–58). Cham.
5. Singh, I., Kaur, A., Agarwal, P., & Idrees, S. M. (2024). Enhancing security and transparency in online voting through blockchain decentralization. *SN Computer Science, 5*(7), 921.
6. Farooq, M. S., Ansari, Z. K., Alvi, A., Rustam, F., Díez, I. D., Mazón, J. L., Rodríguez, C. L., & Ashraf, I. (2024). Blockchain-based transparent and reliable framework for wheat crop supply chain. *PLoS One, 19*(1), e0295036.
7. Madumidha, S., Ranjani, P. S., Vandhana, U., & Venmuhilan, B. (2019). A theoretical implementation: Agriculture-food supply chain management using blockchain technology. *In 2019*

TEQIP III Sponsored International Conference on Microwave Integrated Circuits, Photonics and Wireless Networks (IMICPW) 2019 May 22 (pp. 174–178). IEEE
8. Zhang, X., et al. (2020). Blockchain-based safety management system for the grain supply chain. *Ieee Access, 8*, 36398–36410.
9. Duan, J., Zhang, C., Gong, Y., Brown, S., & Li, Z. (2019). A content-analysis based literature review in blockchain adoption within food supply chain. *International Journal of Environmental Research and Public Health, 17*(5), 1784.
10. Sunny, J., Undralla, N., & Madhusudanan Pillai, V. (2020). Supply chain transparency through blockchain-based traceability: An overview with demonstration. *Computers and Industrial Engineering, 150*, 106895.
11. Galvez, J. F., Mejuto, J., & Simal-Gandara, J. (2018). Future challenges on the use of blockchain for food traceability analysis. *TrAC Trends in Analytical Chemistry, 107*, 222–232.
12. Motevalli-Taher, F., Paydar, M. M., & Emami, S. (2020). Wheat sustainable supply chain network design with forecasted demand by simulation. *Computers and Electronics in Agriculture, 178*, 105763.
13. Nadimi, M., Hawley, E., Liu, J., Hildebrand, K., Sopiwnyk, E., & Paliwal, J. (2023). Enhancing traceability of wheat quality through the supply chain. *Comprehensive Reviews in Food Science and Food Safety, 22*(4), 2495–2522.
14. Mesterházy, Á., Oláh, J., & Popp, J. (2020). Losses in the grain supply chain: Causes and solutions. *Sustainability, 12*(6), 2342.
15. Thakur, K. S., Ahuja, R., & Singh, R. (2024). IoT-GChain: Internet of things-assisted secure and tractable grain supply chain framework leveraging blockchain. *Electronics, 13*(18), 3740.
16. Alkahtani, M., Khalid, Q. S., Jalees, M., Omair, M., Hussain, G., & Pruncu, C. I. (2021). E-agricultural supply chain management coupled with blockchain effect and cooperative strategies. *Sustainability, 13*(2), 816.
17. Hosseini-Motlagh, S. M., Samani, M. R. G., & Abbasi Saadi, F. (2021). Strategic optimization of wheat supply chain network under uncertainty: A real case study. *Operational Research, 21*(3), 1487–1527.
18. Agarwal, U., Rishiwal, V., Shiblee, M., Yadav, M., & Tanwar, S. (2024). Blockchain-based intelligent tracing of food grain crops from production to delivery. *Peer-to-Peer Networking and Applications, 17*(6), 3722–3749.

Blockchain-Based Secured Architecture for E-tendering and Bidding System

Kanika Taneja, Ashish Kumar Mourya, and Sheikh Mohammad Idrees

1 Introduction

Bidding and procurement methods have long been affected by challenges including inefficiency, delays, corruption, and a lack of transparency. Traditional centralized systems are vulnerable to fraudulent activity, partiality, and manipulation, which can lead to lower-than-ideal results and significant financial losses. Current computerized tendering processes are not "proper and transparent" since information is not shared with everybody involved. Disclosure of information occurs "as they please." For example, when a company is selected to win a contract, the other companies that placed bids on the same tender are not notified as to why their proposal was rejected and why the chosen company was selected as the winner [1]. Although a company can ask for this information, getting it requires a lot of work. While these records can be audited, their evaluation takes time. These portals' security is a severe issue that can lead to fraud and data tampering from a centralized database in addition to their lack of transparency [2]. A company could suffer significant financial and strategic losses if a hacker manages to obtain access to this centralized database and leaks bids to rival companies [3]. Blockchain technology can be used

K. Taneja
Department of Computer Science and Engineering-Data Science, ABES Institute of Technology, Ghaziabad, Uttar Pradesh, India

A. K. Mourya (✉)
Department of Information Technology, ABES Institute of Technology, Ghaziabad, Uttar Pradesh, India
e-mail: ashish.morya@abesit.edu.in

S. M. Idrees
Department of Computer Science, Norwegian University of Science and Technology, Gjøvik, Norway

S. M. Idrees et al. (eds.), *Blockchain and Machine Learning Innovations*, Signals and Communication Technology,
https://doi.org/10.1007/978-3-031-88095-7_6

to address these security issues because of its distinctive block-based architecture for transaction management and string emphasis on information decentralization. Thus, blockchain technology and smart contracts may be used to create a transparent, decentralized, and safe bidding framework [4]. This would give bides easier control over the tender portal's operations and allow them to keep an eye on everything it does. In the world of electronic auctions today, bids are typically divided into two categories: sealed bids and public bids. In public auctions, bidders can increase their bids gradually until no bidder is willing to bid more, and the highest bid wins the product. Another name for these auctions is multi-bid auctions. In sealed bids, bidders submit their bid once encrypted, and the auctioneer compares all bids to determine the highest bid at the conclusion of the auction. Participants are only permitted to place one bid at a time in single-bid auction [5]. Unfortunately, collusion is a significant disadvantage of electronic seal bids. Prior to the bids being opened, bidders frequently communicate with auctioneers through a centralized middleman in order to get around this, which raises privacy issues and transaction fees. They frequently communicate with auctioneers through a centralized middleman in order to get around this, which raises privacy issues and transaction fees. In order to solve these problems, this study suggests using blockchain technology in e-auction.

1.1 Public Blockchain

A decentralized network with transparent, immutable, and cryptographically secured transactions powers a public blockchain [6]. For e-tendering and building processes, public blockchain gives the number of benefits (Fig. 1). By strengthening all the stakeholders with real-time access to the entire process, they first ensure transparency and flexibility by enabling them to validate bids and check their progress. Second, data integrity is ensured and details of bid and terms of contract

Fig. 1 Public blockchain

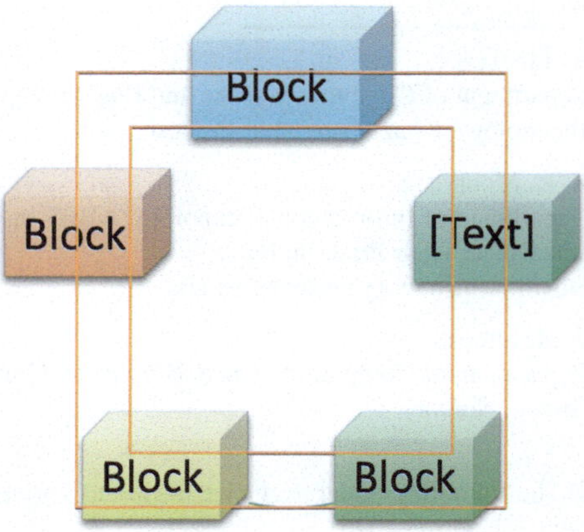

cannot be tampered or altered with due to the unchangeable nature of public blockchain. The result has considerably increased trust and condensed the likelihood of manipulation or fraud [7].

As we know that the public blockchains operate exclusive of central authority, these chains also promoted trustlessness and decentralization. As no single party controls the development, the chance of partiality or bias is minimized. Furthermore, these public chains maintain smart contracts, which mechanize various steps in the e-tendering process. Smart contracts accelerate procedures and lessen manual labor via automatically enforcing the rules and conditions of contracts. Moreover, by employing advanced crypto-graphic algorithms, public chains improve privacy and security [8]. Participants can be assured their data remains protected with encrypted and cryptographically secure transactions [9]. All these components in any organizations can increase the integrity, effectiveness, and reliability of their processes of procurement. All the collaborative parties gain from fair bidding and competitive conditions accessible by public blockchain, which also promote modernism in different procurement practices.

1.2 Smart Contract

Smart contract means any digitally signed document and also has the ability to perform tasks automatically or, we can say, autonomously. The main characteristics of this contract are that it is unchangeable, secure, and transparent due to its occurrence through blockchain [10–12]. These contracts are also currently used in procurement and e-tendering for bidding, where the process has changed dramatically. Through automation procedures like bid execution, evaluation, and submission, the need for middlemen has been eliminated, reducing the likelihood of manipulation or fraud [13]. Smart contracts give permission to authorized entities to communicate freely, also guaranteeing that their duties are performed effectively and transparently. The deployment process of smart contracts via blockchain in e-tendering and bidding systems has the capacity to increase the procurement process's effectiveness, integrity, and transparency, which benefits all stakeholders involved in the shared network. Firms and organizations make use of a pre-made smart block to create a tender.

The details of project, the deadline of bid, and the criteria of selection are all outlined in this contract. The technology of blockchain provides bidder with immediate access to the data and information of tender. In order to submit their bids, bidders execute the smart contract. Bids can be encrypted for added security, and the organization can only view them after the deadline. The smart contract serves as a reminder as the deadline approaches, immediately ending new bid submissions. The winning bidder will have access to project details after the smart contract is configured to evaluate bids based on present criteria.

To define a tender, organizations use a prepared smart contract. In the contract, details about the due dates, project specifications, and other tender information are included in this contract. Using smart contracts in the bidding process has a number of benefits, including:

- Transparency: It stores all bids and tender data based on blockchain technology fosters an environment that is both transparent and equitable.
- Efficiency: Automation expedites the process by lowering human labor and saving people's time.
- Trust: Smart contracts build a foundation of trust between the company and bidders by doing away with the need for an intermediary.

2 Existing System

The current tendering processes in India are vulnerable and inefficient to fraud because it is heavily centralized and dependent on paper based and manual processes. The procedure involves centralized data storage, physical submission of bids, manual committee evaluation of bids, and, all of which leads to delays, a lack of transparency, and opportunities for corruption. The existence of multiple middlemen and asymmetry of knowledge enable bid rigging, favoritism and bribery. It is challenging to hold people investigate, accountable, and inconsistencies when there is no permanent record keeping. The incorrect structure undermines public trust and often ends in less than financial losses and deal outcomes. An e-tendering solution that is transparent, robust, and up to date is urgently needed because the current system is unable to keep up with the digital revolution due to its reliance on practices which are not relevant.

3 Proposed System

The recommended solution to streamline and fasten the entire tendering process is smart contracts. It is a decentralized platform based on blockchain technology. It consists of a chain of nodes on a blockchain that are monitored by authorized entities, including government agencies, public sector enterprises, and auditors which work independently and can give transparency and protection from manipulation which is ensured by encoding evaluation standards, in terms of contracts and tender laws into smart contracts. The encrypted bids are submitted by bidders via a secure interface, and by anti-collision algorithms, we can prevent bid rigging. The unchangeable blockchain ledger maintains an extensive trail of audits that enables grievance resolution and effective monitoring. Economical and ethical public procurement is motivated by integration with the current system, scalable, compliance with protocols, standards, rules, regulations, and stakeholders' engagement tactics that enables smooth procurement across various different sectors.

4 Research Gap

Traditional electronic tendering systems face limitations in transparency, data integrity, audibility, and efficiency due to their centralized architecture, lack of immutable audit trails, and involvement of intermediaries. These shortcomings open opportunities for manipulation, disputes, delays, and unethical practices.

The distributed, decentralized, and cryptographically secure nature of blockchain gives a prominent solution to solve some of the gaps of Blockchain research area. Some of the important blockchain's inherent properties like transparency, immutability, and smart contracts, the platform like blockchain-based e-tendering platform provide:

- A decentralized system without single points of failure, reducing data tampering risks
- Immutable and auditable records of all tendering activities on a distributed ledger
- Automated execution of tendering rules and evaluation criteria through smart contracts
- Secure and verifiable bid submissions with prevention of bid rigging
- Minimized intermediaries, reducing delays and opportunities for corruption

Addressing these research gaps through blockchain can foster trust, transparency, cost-efficiency, and ethical practices in tendering processes across industries and government sectors.

5 Research Objectives

The primary objective of this research is to evaluate and develop a comprehensive blockchain-based e-bidding and tendering system that addresses the disadvantages of traditional centralized systems and addresses the unique properties of this immutable technology. Particularly, the research aims to accomplish the following objectives:

i. Design a decentralized and secure architecture for a blockchain-based e-tendering platform that assures transparency, data integrity, and auditability across the full tendering lifecycle, from tender advertisement to bid submission, assessment, and contract award.
ii. Development of architecture of smart contract that operates automatically, removing the intermediaries, the needs and minimizing the opportunity for changes or human mistake.
iii. Using cryptographic protocols and anti-collision algorithms provide safe and verifiable procedures for submitting bids, guaranteeing secrecy, and discouraging bid ragging or bidder collusion.
iv. Provides a structure of good governance that will give all the access restrictions, obligations, and responsibilities for all parties included in distributed e-bidding

system, such as higher authorities, management, bidders, other regulatory agencies and different auditors which are independent in nature.

v. Providing assurance that the model is ready for world-wide implementation across various sectors and industries do scalability testing and performance to handle large volumes of transactions, various concurrent users, and comprehensive procedures of tendering.

vi. Conduction of comprehensive risk-mitigation plans and deep external security audits, you may check the security of system and protection against possible threats including flaws in smart contracts, data breaches, and cyber attacks.

By achieving these goals, this research development make a scalable, strong, and secure e-bidding and tendering system which has the capacity to change public procurement procedures, increased transparency, less costs, and motivate moral business conduct in a variety of industry and government sectors.

The main objective is to achieve a distributed and decentralized safe and transparent e-tendering platform which uses blockchain technology and smart contracts. The primary objective includes the automation of the process of bidding with the assurance of immutability and smart contracts data auditability with blockchain implementing encrypted and secure bid submission with seamless integration of procurement systems and anti-collision safeguards. The ultimate objective is to provide a robust blockchain system that enhances public procedure across industries, lowers the risk of corruption, promotes the ethical conducts, and maximizes the cost-effectiveness.

The three main stages of the blockchain-based e-tendering and bidding process are bid evaluation, bid submission, and tender creation. During these stages, tender methodically started, bids are safely submitted over blockchain platform [14, 15], and bids are impartially accessed utilizing smart contracts and decentralized system.

The creation of the tender, the submission of the bid, and the assessment of the bid are three primary points in the blockchain-based e-tendering and bidding process. Tenders must be systematically initiated throughout these phase, bids must be securely submitted using the blockchain platform, and bids must be objectively evaluated using smart contracts and decentralized procedure [16, 17]; the concise approach ensures a thorough examination of the e-tendering process inside the blockchain framework while maintaining the authenticity and originality.

6 Implementation

To implement the blockchain-based electronic tendering and bidding system, a decentralized application (DApp) utilizing the Ethereum blockchain and smart contract must be created. The system's frontend is built using the next.js react framework, which provides administrators, bidders, and tender issuers worth an easy-to-use interface. The backend smart contract are created in solidity and then deployed on the network using tools like Hardhat. These smart contracts encode the evaluation criteria and tendering rules and automate the bidding and assessment processes. The DApp connects to the blockchain using web3.js, a JavaScript tool for establishing a connection with Ethereum nodes. Strong security features like encryption for bid submissions and anti-collusion strategies are also included in the system. The deployment also includes integration with w-procurement platforms and existing corporate systems.

This study suggests a novel decentralized application (DApp) that utilizes the special features of blockchain technology in response to the urgent need to address the numerous issues besetting conventional tendering and bidding procedures. The proposed system, termed "Blockchain based E-tendering and Bidding System, i.e., BlockBid," is a public permissioned blockchain network that facilitates transparent, auditable, and efficient tendering across various industries and government sectors.

The process starts when the authority responsible for issuing tenders accesses the e-tendering platform and creates a new tender by defining the parameters of the project, the standards of assessment, and the due dates. The blockchain network uses smart contracts that have this data recorded in it. The portal is then accessed by registered bidders, who study the released tender and submit their encrypted offers, which are safely stored on the blockchain. Following the bid deadline, the authority initiates the smart contract-managed automated review procedure, which evaluates the submitted bids impartially following the predetermined standards. The results of the review, including the successful bidder, are entered into the blockchain ledger as an unchangeable transaction. Through the e-tendering platform, bidders and the authority may both examine and verify the outcomes. Next actions, like signing a contract.

The workflow diagram illustrates the sequence of steps involved in the e-tendering process on the blockchain platform. It begins with the tender issuing authority initiating a new tender by specifying project details, evaluation criteria, and deadlines, which are encoded into smart contracts on the blockchain. Registered bidders then review the published tender and submit encrypted bids securely recorded on the blockchain. After the bidding deadline, the authority triggers the automated evaluation process governed by smart contracts, assessing bids objectively based on predefined criteria. The evaluation outcome, including the winning bidder, is immutably recorded on the blockchain ledger, accessible to all stakeholders through the e-tendering platform.

The workflow is shown in (Fig. 2) which consists of various activities as mentioned below:

i. The application of e-tendering system starts with registration of tenderer authority and bidders, after the registration the tenderer will be redirected to the tenderer dashboard and bidder will be redirected to the bidder dashboard.
ii. The tenderer will create or upload the tender to the blockchain network. On the other hand, the bidder will be able to view and submit their bid according to the requirement of the tenderer.
iii. The submitted bid will be verified by the smart contract. After verifying the bids, the best bids will be selected using Selection Algorithm Using Single

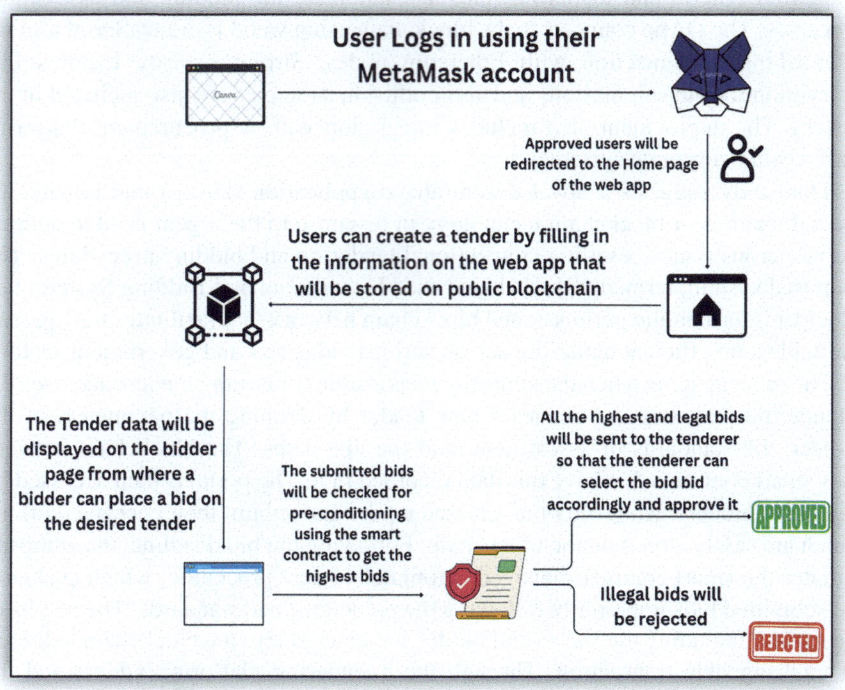

Fig. 2 Proposed work flow

Reservation Mechanism after which the winning bids will be sent to the tenderer for further negotiations.

Block Bid is a consortium network of blockchain in which authorized participants—procuring authorities, regulatory agencies, registered bidders, and independent auditors—operate all the nodes. This hybrid architecture maintains an immutable and distributed record of transactions while ensuring that only authorized entities are allowed to participate in the processes of tendering. The core elements of Block Bid are a collection of smart contracts that encapsulates and automate the entire process of tendering, from the promotion of tender and its documentation till submission, evaluation, and awarding of contracts. These self-executing contracts remove the need for middlemen, which lowers the efforts of administration and the possibility of human manipulation and error.

Block Bid's most predominant features are:

i. Transparent and Unchangeable Tender Documentation: All specifications of project, requirements of eligibility, and criteria of evaluation, terms and conditions, and information of tender are stored on the network of blockchain using smart contracts, which establish an easily reachable, single source of truth for all network participants with consensus.

ii. Verifiable and Secure Bid Submission: Using the web interface of Block Bid, which is safe by robust cryptographic techniques, various bidders can submit bids which are encrypted to guarantee bid which is confidential in nature until the designated opening time. Anti-collusion algorithms prevent and detect bid-rigging and bidder cooperation.
iii. Automated and Auditable Bid Evaluation: The smart contracts remove any suspicion of discrimination or bias by automatically evaluating bids submitted in line with present criteria. The unchangeable blockchain database meticulously records each step of the review process and each decision.
iv. Effective Contract Award and Execution: Smart contract manages all outstanding transactions and agreements after the completion of review process, cutting down delays and problem with automated winning offers.

Deep simulations and test installations have been analyzed to check the system's performance, usability, consistency, and precautions in the range of tendering situations. Compliance with applicable regulations have been fixed by the assurance of independent auditors and authorized regulatory entities handling the whole tendering process in situations of real time. The design of Block Bid has developed with interoperability, scalability, and simplicity. Its main objectives are utilizing the technology of blockchain and standards. A decentralize grievance resolution method gives transparency in all issues settlement.

This chapter provides a comprehensive working and review of the Bid system, which includes details on its architectural components and regulatory protocols. The outcomes of our finding and testing will definitely fasten up the e-bidding with minimum cost. We assured that our research will also promote the moral conduct and increase the trust of participating organizations in public procurement. Next.js, which is based on the react framework, has been used for the development of the frontend. Its server-side rendering capabilities and performance optimization abilities make it a user-friendly e-trading platform with efficient routing protocols. Hardhat, an application toolkit for the Ethereum development environment, is used for writing and deploying smart contracts. Our first natural choice for the implementation of core smart contracts was solidity (Ethereum's contract-oriented language) that governed and automated our tendering process on the blockchain.

7 Result and Discussion

Our study examined the functioning and deployment of blockchain-based e-tendering and bidding systems by comparing them with conventional tendering methods and increased security. Comparing the tamper-proof features and security provided by Ethereum-based e-tendering platforms with the disadvantages found in the existing tendering systems was our main goal. Our proposed study findings clearly give that e-tendering and bidding system based on technology of blockchain gives better results than traditional approaches. After a thorough evaluation and analysis, we have determined a number of potential and important factors that show how technology of blockchain may change the procedures of bidding.

i. Security Enhancement: By digitally encrypting transactions and sensitive data, blockchain-based tendering systems provides better security features. By ensuring that contract details and bid are stored safely and cannot be changed with the Blockchain's distributed structure which reduces the risk of data breaches and unauthorized access.
ii. Immutable and Accurate Documentation: Because technology of blockchain which is immutable in nature, records associated to tenders are assured to be immutable and accurate. An unbreakable and impenetrable continuous chain of records is developed by safely connecting each transaction to every another transaction. The confidentiality of the e-tendering process is supported by this unbreakable system of recordkeeping.
iii. Reduction in Third-Party Involvement: Blockchain-based e-tendering system lowers the challenges of administration and streamlines the process of procurement by doing away with the requirement for intermediaries or other third parties. Based on the technology of blockchain, smart contracts automated the contract enforcement and eliminate the need for trustworthy middlemen, enabling transactions at peer-to-peer level at a reduced cost.
iv. Prejudice Elimination: Blockchain based E-tendering virtual environment uses distributed, transparent, decentralized procedures of consensus to promote no-bias environment and safe environment in the procurement process. The equal access for submissions of bid and details of tender is provided to all interested parties which helps in the promotion of fast competition and lowers the possibility of corruption or bias.

7.1 Tender Block Creation

Every time a tender or new bid or tender is created in an electronic bidding and tendering system built on the technology of blockchain, a new block is to be added in the network of blockchain. Designated as an entry in the digital ledger, the particular block holds all information about the tender and relevant bid in the form of a unique hash code. The tender ID, details of bidder, amount of bid, timestamp of bid, are all safely represented by this hash code, which also serves as a safe and secure identity.

Throughout the e-bidding and tendering process, the technology of blockchain guarantees integrity, transparency, and tamper-proof record-keeping by storing this data in an immutable decentralized manner. The confidentiality of the tendering process is applied with each new block added to the structure of blockchain, which leaves an undisturbed trail of transactions that can be efficiently viewed and verified by user of the network in blockchain.

The creation of Bid in the blockchain e-tendering system is given by particular smart contracts which are implemented on the network of Ethereum. A smart contract is instantiated with the standards, specifications, and deadlines of a new tender upon its publication. Various bidders test the data which is stored on blockchain network by logging into this proposed platform. In order to submit bids, various bidders communicate with the smart contract using the interface of the platform, sending encrypted bid data including requirements, price, and required documentation. An immutable audit trail is produced by the smart contract, which stores the bid which is encrypted, the address of bidder, and the timestamp on the Ethereum blockchain. This procedure continues until the deadline for bids, at which point the smart contract denies the submission of further bids. A tamper-proof and safe, bidding procedure is ensured by the Blockchain's automatic execution and transparency of smart contracts, which removes any mistakes and interference of humans.

7.2 Deployment and Application

The final deployment of the front end of a blockchain-based e-bidding and tendering system with Hardhat, Tailwind CSS, and Next.js requires many potential stages to ensure a fast and smooth operation. Firstly, the framework of Next.js is used to provide strong support for the creation of server-side rendering and static websites, which improves speed and SEO. Because Tailwind CSS gives priority to function above style, it is accumulated to enable responsive and quick creation of user interface. Firstly, a setup of Next.js project must be fixed, and then Tailwind CSS needs to be configured for style. In order to guarantee the effectiveness and security of the backend logic for tendering and bidding, Hardhat is utilized in synchronization with Ethereum blockchain smart contract deployment and management. The front end is assured to be secure and safe upon final deployment to a platform such as Vercel and Netlify.

The dashboard of tender creation is an interface which is user friendly and allows private organizations, government organizations, and other authorized entities to initiate and give information about new tenders on the network of blockchain. It provides a seamless experience for giving tender details which includes project requirements, criteria of evaluation, deadlines of submission, and required documentation. The dashboard leverages Tailwind CSS and Next.js to deliver a nice and visually appealing usability and design, across different screen sizes and devices. Upon the completion of the tender creation process, the dashboard connects with the backend to deploy the corresponding smart contract on the Ethereum blockchain, ensuring immutability, decentralized access and transparency, to the information of tender.

Using Tailwind CSS with Next.js, the front end of the bid generation dashboard in a blockchain-based e-bidding and tendering system provides an experience which is user-friendly for submission of bid. Tailwind CSS offers an appealing and responsive design, while Next.js delivers seamless navigation with quick rendering.

Through dynamic forms that command to check data in real-time scenario, users will try to submit bid details which will trigger smooth communication with the blockchain using Ethers.js or Web3.js integration with smart contracts which ensures unchangeable and safe bid submissions. This system management ensures that bidders will have an effective experience and easy-to-use strategy while upholding the bidding process's transparency and integrity.

8 Conclusion

Through the detailed investigation and hands-on simulation, it shows that the ethereum-based e-bidding and tendering system works better than traditional centralized platforms in terms of advantages and functionality. Our recommended approach implements an issue which is very prominent with public contacting and procurement procedure by utilizing the important component of ethereum smart contracts and the blockchain technology. Throughout the whole binding process, the distributed and decentralized ethereum blockchain gives the best standard of data transparency, integrity, and protection against manipulation. The cryptographically, immutable and secure ledger kept updated by the ethereum blockchain network gives an impenetrable and auditable record of all operations of tendering in comparison to the centralized database that is prone to possible data manipulation and single point of failure.

In addition to this, the ethereum blockchain smart contracts implementation gives us the scope to enforce standards and in achieving the automation, contractual obligations, and tendering process by eliminating the need for the middleman, and these self-executing contracts narrows the possibility of unlawful behavior, prejudices, and human mistakes. Our thorough performance through simulation and scalability testing successfully demonstrated the capacity of system to manage high transactional volumes, complications of tendering process and multiple users at once confirming its rightness for practical implementation across various industries and sectors. In addition to this, our solutions provide strong security assessments technique which provides protection against various dangerous threats like flaws in smart contract, data breaches and cybersecurity. The ethereum-based e-bidding and tendering platform has potentially lowered the cost, increased efficiency and potential stakeholders' confidence in processes through trials and hits.

Additionally, the ethereum blockchain inherent transparency and auditability have significantly improved the public confidence in the bidding processes' fairness. The blockchain ensures the accountability and expenditures grievance redressal process by providing immutable records that are accessible and verifiable to numerous parties including independent auditors and regulatory organizations.

In conclusion, our study shows how ethereum-based e-tendering platforms have the potential to transform public procurement procedures and lower the risk connected with conventional approaches. Our solution delivers unmatched security, transparency, affordability, and resistance to manipulation through the use of smart contracts and blockchain technology. This makes it possible for a number of government and commercial sectors to implement component, moral, and effective tendering processes.

9 Future Scope

Although the implementation and efficacy of blockchain-based e-tendering and bidding systems have been better understood, thanks to this research; there are a number of directions that merit further investigation.

- **Minimization of Transactional Costs**: Researching strategies to reduce transactional costs related to blockchain-based e-tendering system is a promising direction for future work. In order to lessen the financial burden in users and improve cost-effectiveness, future research could concentrate on optimizing gas and transaction fees on Ethereum-based platforms.
- **Integration of Hyperledger Fabric**: In permissioned blockchain framework including Hyperledger Fabric, the possible future research can be done in e-tendering and bidding systems. The Hyperledger Fabric modular design and its support for permissioned and private networks provide special advantages, such as the potential to completely eliminate the expenses of gas. In future, we can look into the feasibility and benefits of utilizing Hyperledger Fabric to create affordable and adjustable e-tendering procedures.
- **Enhanced Privacy and Confidentiality**: In future, we can look into various ways to improve confidentiality and integrity on blockchain-based e-tendering systems, especially in industrial sector and government procurement procedures. To retain transparency and audibility, we try to save the data which is sensitive, innovations like zero-knowledge proofs and privacy-preserving smart contracts could be studied.
- **Interoperability and Integration**: To enable smooth integration and resolve interoperability issues between blockchain-based e-tendering systems and current procurement infrastructure, more research is required. In order to facilitate interoperability between various legacy systems and blockchain platform, research of future could concentrate on creating protocols and standards. This would boost productivity and streamline procurement procedures.
- **Scalability and Performance Optimization**: Future studies should focus on the scalability and performance optimization issues related to e-tendering and bidding systems as blockchain technology develops. In order to increase throughput and handle increasing transaction volumes without compromising security or decentralization, research could investigate novel consensus mechanisms, sharding, and off-chain solutions.

References

1. Pandian, A. P., Palanisamy, R., & Ntalianis, K. (Eds.). (2021). *Proceedings of International Conference on Intelligent Computing, Information and Control Systems: ICICCS 2020*. Springer.
2. Idrees, S. M., Nowostawski, M., Jameel, R., & Mourya, A. K. (2021). Security aspects of blockchain technology intended for industrial applications. *Electronics, 10*(8), 951.
3. Ramazhamba, P. T., & Venter, H. S. (2022). A distributed model for sharing tendering project information in the South African Local Government. In *Proceedings—2022 International Conference on Cyberworlds, CW 2022* (pp. 243–250). https://doi.org/10.1109/CW55638.2022.00055
4. Goswami, Y., Agrawal, A., & Bhatia, A. (2020). E-governance: A tendering framework using blockchain with active participation of citizens. In *International Symposium on Advanced Networks and Telecommunication Systems, ANTS, 2020-December*. https://doi.org/10.1109/ANTS50601.2020.9342816
5. Taiwan zhishichuangxinxuehui. (n.d.). *Applied System Innovation: Proceedings of 4th IEEE International Conference on Applied System Innovation 2018 (IEEE ICASI 2018)*. Chiba, Japan, April 13–17, 2018.
6. Hassija, V., Chamola, V., Krishna, D. N. G., Kumar, N., & Guizani, M. (2021). A blockchain and edge-computing-based secure framework for government tender allocation. *IEEE Internet of Things Journal, 8*(4), 2409–2418. https://doi.org/10.1109/JIOT.2020.3027070
7. Mourya, A. K., & Idrees, S. M. (2020). Cloud computing-based approach for accessing electronic health record for healthcare sector. In *Microservices in big data analytics: Second international, ICETCE 2019, Rajasthan, India, February 1st-2nd 2019, revised selected papers* (pp. 179–188). Springer Singapore.
8. Sai Reddy, E., Sathvik, M., Rajaram, V., & Rao, C. P. (2023). An intelligent tender management system using block chain and IPFS. In *International Conference on Sustainable Computing and Smart Systems, ICSCSS 2023—Proceedings* (pp. 1497–1502). https://doi.org/10.1109/ICSCSS57650.2023.10169649
9. Kapil, G., Kumar, N., Mourya, A. K., & Kumar, V. (2025). Securing big healthcare data using attribute and honey-based encryption in cloud environment. *The Journal of Supercomputing, 81*(1), 181.
10. Diadia, P., Tamgno, J. K., & Kora, A. D. (2022). Dematerialization of public procurement approach based on hyperledger fabric blockchain using OCDS. In *Proceedings—2022 IEEE International Conference on e-Business Engineering, ICEBE 2022* (pp. 184–189). https://doi.org/10.1109/ICEBE55470.2022.00040
11. Satybaldy, A., Subedi, A., & Idrees, S. M. (2024). Decentralized key management for digital identity wallets. In *Blockchain transformations: Navigating the decentralized protocols era* (pp. 47–58). Cham.
12. Singh, I., Kaur, A., Agarwal, P., & Idrees, S. M. (2024). Enhancing security and transparency in online voting through blockchain decentralization. *SN Computer Science, 5*(7), 921.
13. Idrees, S. M., Nowostawski, M., Jameel, R., & Mourya, A. K. (2021). Privacy-preserving infrastructure for health information systems. In *Data protection and privacy in healthcare* (pp. 109–129). CRC Press.
14. Marati, V., Pinagani, S., Sathwika, N., Lakshmi, V., & Yasaswini, A. (2023). Revolutionizing tender management through blockchain technology. In *6th International Conference on Inventive Computation Technologies, ICICT 2023—Proceedings* (pp. 1172–1177). https://doi.org/10.1109/ICICT57646.2023.10134275
15. Yutia, S. N., & Fathiyana, R. Z. (2022). Performance analysis of blockchain in tendering process based on hyperledger framework. In *2022 10th International Conference on Information and Communication Technology, ICoICT 2022* (pp. 190–193). https://doi.org/10.1109/ICoICT55009.2022.9914877

16. Li, X. C. (2020). BCES: A blockchain based credible E-bidding system. In *2020 IEEE 6th International Conference on Computer and Communications, ICCC 2020* (pp. 2415–2421). https://doi.org/10.1109/ICCC51575.2020.9345176
17. Bhole, K., Galphat, Y., Bathija, K., Saini, T., & Bhosale, A. (2021, January 15). Allocation and tracking of public funds using blockchain. In *2021 International Conference on Nascent Technologies in Engineering, ICNET 2021—Proceedings*. https://doi.org/10.1109/ICNTE51185.2021.9487782

AI-Powered Predictive Analytics in Decision-Making in Smart Cities

Varsha Bhatia and Sunita Kumawat

1 Introduction

The concept of smart cities is gaining its place in today's increasingly urbanized world, offering a solution to the demanding challenges of climate change and efficient urban services. This chapter digs into the critical role of AI-powered predictive analytics in shaping the future of smart cities. Smart cities are revolutionizing urban development by harnessing cutting-edge technologies to improve the lives of their residents and optimize resource utilization. At their heart, smart cities blend information and communication technologies (ICT) with urban infrastructure, creating a network of interconnected devices that collect and analyze real-time data [1]. This data-driven approach empowers cities to respond swiftly and effectively to challenges like traffic congestion and energy consumption. The smart city concept spans a wide range of applications, from intelligent transportation systems and smart grids to efficient waste management and citizen engagement platforms [2]. By connecting these diverse systems through IoT devices and sensors, smart cities create a comprehensive digital ecosystem.

As the cities grow increasingly complex, traditional methods of urban planning and management are crippling. Data-driven decision-making has emerged as a powerful tool to navigate this complexity. By harnessing the vast amounts of data generated by urban life, city officials and planners can gain unprecedented insights into how cities function, understand the needs of their citizens, and even predict future trends [3].

V. Bhatia
ADYPU Pune, Pune, Maharashtra, India

S. Kumawat (✉)
Amity University Haryana, Gurgaon, Haryana, India
e-mail: skumavat@ggn.amity.edu

S. M. Idrees et al. (eds.), *Blockchain and Machine Learning Innovations*, Signals and Communication Technology,
https://doi.org/10.1007/978-3-031-88095-7_7

This shift toward data-driven governance allows for more precise, timely, and effective solutions to urban challenges. For example, by using sensor data to predict infrastructure failures, cities can proactively schedule maintenance and prevent costly breakdowns [4]. Real-time traffic data can be used to optimize traffic flow and reduce congestion [5]. In addition, data-driven approaches promote transparency and accountability in governance, as decisions are based on evidence and measurable results.

Artificial intelligence (AI) and machine learning (ML) are at the forefront of smart city innovation. AI enables the processing of complex urban data, while ML systems learn and improve over time, driving applications like predictive policing and healthcare services. Blockchain technology, though still in its early stages, offers a secure and transparent way to manage urban data and transactions through smart contracts. Together, these technologies are transforming cities into more efficient, sustainable, and citizen-focused environments.

This paper is divided into five sections. Section "Introduction" introduces emerging technologies, section "Smart City Applications" explores smart city applications, section "Use Case: Deep Learning Implementation in Traffic Management" provides a brief overview of traffic management use cases and outlines a coding framework. Section "Case Studies" presents four case studies. Section "Implementation for AI/ML" delves into the technical implementation of AI/ML and associated challenges. Finally, section "Discussions" discusses future considerations, followed by a conclusion and references.

1.1 Predictive Analytics in Smart Cities

Predictive analytics has become a foundation of smart city initiatives, providing invaluable insights into urban dynamics and empowering proactive decision-making. Predictive analytics is a tool which utilizes historical data, statistical models, and machine learning techniques to predict future outcomes [6].

Predictive analytics empowers city planners and administrators to anticipate challenges before they strike, enabling them to take proactive steps. By optimizing resource allocation, enhancing decision-making, and improving long-term planning and policy development, predictive analytics helps cities transition from reactive problem-solving to proactive urban management. This ultimately leads to more efficient, sustainable, and liveable cities.

Predictive models in smart cities typically follow a multi-step process, including data collection, data cleaning, model selection, training, validation, prediction, and interpretation. Common techniques used in predictive modeling include time series analysis, regression models, classification algorithms, and clustering techniques [7]. These models can be continuously updated with new data, allowing them to adapt to changing urban conditions and improve their accuracy over time [8]. The steps involved in the process of predictive analytics are depicted in Fig. 1 given below.

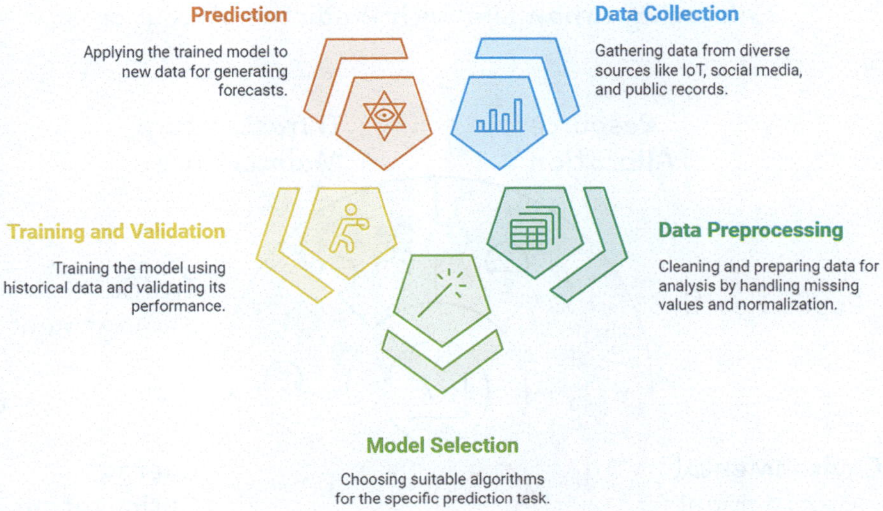

Fig. 1 Steps for performing predictive analytics

Predictive analytics has a wide range of applications in smart cities, benefiting various aspects of urban management. These applications include infrastructure maintenance, traffic management, energy optimization, public safety, urban planning, environmental management, public health, resource allocation, citizen services, and economic development [9] By leveraging these insights, city administrators can make more informed, data-driven decisions, reducing reliance on guesswork or outdated information.

As predictive analytics technologies continue to evolve, their integration into smart city operations is expected to deepen, offering even more sophisticated and accurate forecasting capabilities to address the complex challenges faced by modern cities. Figure 2 shows the various domains in smart cities where predictive analytics can enhance urban lives.

1.2 Artificial Intelligence (AI) and Machine Learning for Urban Systems

Providing effective means of analyzing the complex systems of the city, AI and ML technologies are changing the course of events in urban management [10].

Marvin Minsky and John McCarthy's definition of AI is analogical to any regime that uses a computer program or machine to do acts which evidently demand intelligence. They understand AI systems as those engaging plannings, learnings, reasoning, problem-solving, social intelligence, and creativity which are attributed to intelligence of a human being [11].

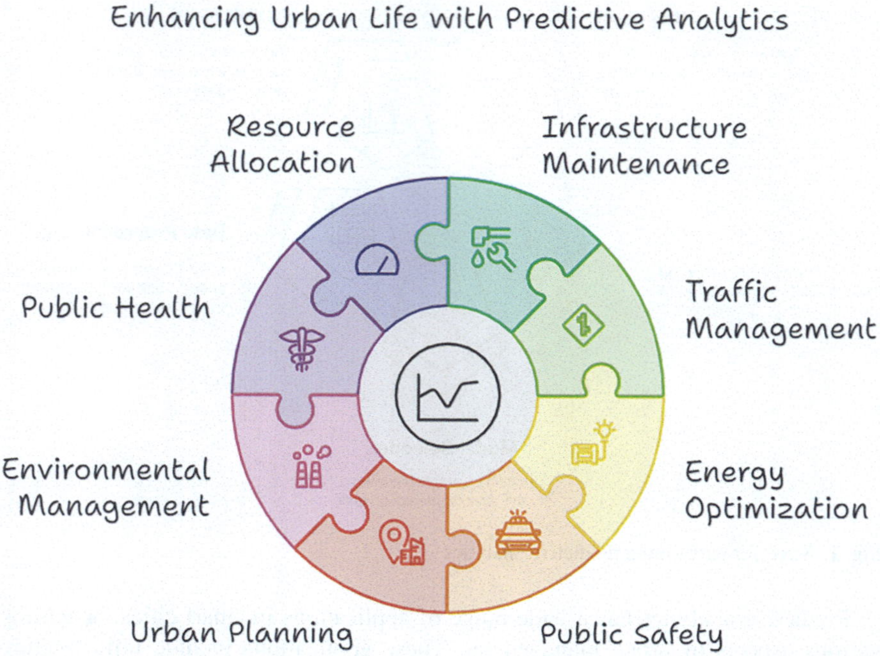

Fig. 2 Predictive analytics in smart city

Deep learning, particularly in machine learning, has been one aspect that has contributed to the recent resurgence of AI. This expansion is further fuelled by the exponential explosion of data which improves the learning of machine learning algorithms. AI has led to development of many interesting opportunities like self-driving cars and delivery robots, but it has become the origin of numerous ethical Issues as well [12].

In smart cities, AI is the general concept of machines performing intelligence-related functions, whereas ML is the subsection of AI that aims on those algorithms that self-enhance with experience.

Incorporating AI and ML in smart cities has enabled great new opportunities for urban transformation and improvement. Such analytical tools as these are to change the way in which the cities operate ushering in a new age characterized by efficiency, sustainability, and quality of life.

Figure 3 depicts how AI/ML applications in urban transformation vary by automation and impact across four key categories.

The union of AI and ML in smart cities aids:

- Processing and analysis of enormous urban data in real time
- Identification of patterns and trends that are unnoticeable to human analysts
- Automation of routine tasks, releasing human resources for more complex tasks
- Adaptive systems that learn and improve over time, enhancing city operations

These advanced technologies can achieve new levels of efficiency, preparing for a future where citizen's wellbeing will be at the forefront [13].

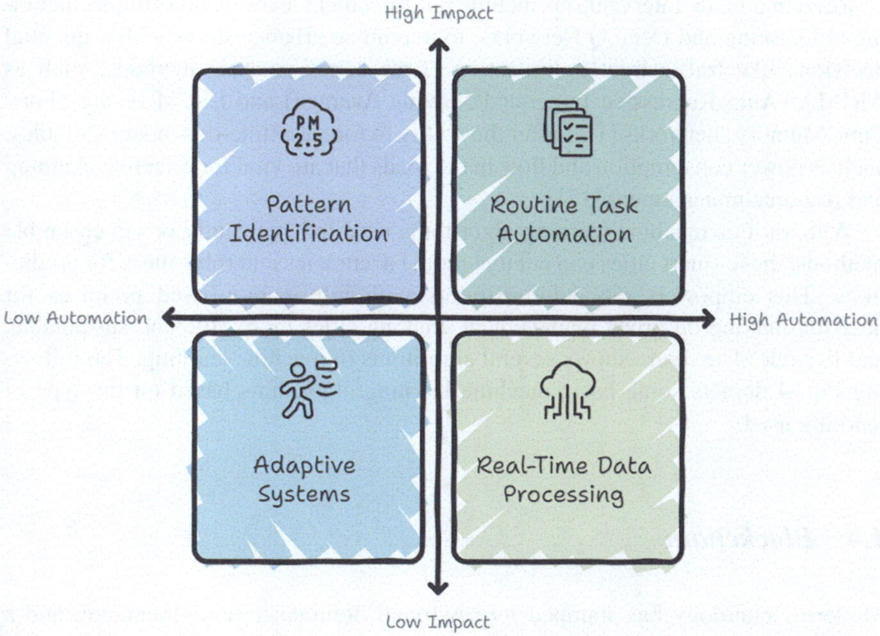

Fig. 3 AI/ML in urban transformation

1.3 Machine Learning Algorithms for Predictive Analytics

The application of machine learning algorithms is very significant in predictive analytics in smart cities. With such intelligent computational techniques, the decision-makers can use the data produced consistently to prepare insights in the urban environments [14].

Several algorithms are highly useful within specialized learning, including supervised learning. An ordinary linear regression is used to solve problems where numerical values are predicted, while logistic regression predicts outputs with binary alternatives. Decision trees and random forests usually provide very efficient classification and regression algorithms and provide interpretable results. Support vector machines (SVM) work very advantageous when both complex and high-dimensional data merging in various predictions. Neural networks can battle very well with complex prediction tasks with their powerful pattern recognition ability [15].

Within the genre of unsupervised learning, clustering algorithms such as K-means assist in discovering useful groupings or segments in urban data. Principal component analysis (PCA) plays a tremendous role in reducing dimensions and extracting features that help to discern an anomaly hidden in the multi-dimensional data.

Reviving, many interventions include reinforcement learning algorithms, including Q Learning and Deep Q Networks, to maximize urban systems with sequential decisions like traffic light optimization. Time series analysis methods, such as ARIMA (Auto Regressive Integrated Moving Average) and LSTM (Long Short-Time Memory) networks [16], offer the ability to forecast time-dependent variables, such as power consumption and flow in the roads that are vital for effective planning and resource management [17].

With various machine learning algorithms used independently or via ensemble methods, those smart cities can enlarge model accuracies and robustness for predictions. This empowers city administrators with information-based premises for decision-making on urban management, making cities more efficient, sustainable, and liveable. The figure shows several algorithms of machine learning. The following Fig. 4 depicts some basic machine learning algorithms based on the type of learning used.

1.4 Blockchain

Modern technology has stamped its newfound dominance, enlightenment, and a lifestyle of its own in every industry; the capability of data handling with proper security measures. These technologies deliver a network construction that provides insights and solutions concerning all inadequacies existent in traditional and centralized systems.

Blockchain works essentially by a network of nodes linked to each other directly to record every interaction with timestamping [18]. Consequently, allowing transactions to be executed transparently and without intermediaries while enforcing security measures. The principle is to build data in blocks in such a way that they form a chain so that it becomes the most difficult to tamper. Once transactions are

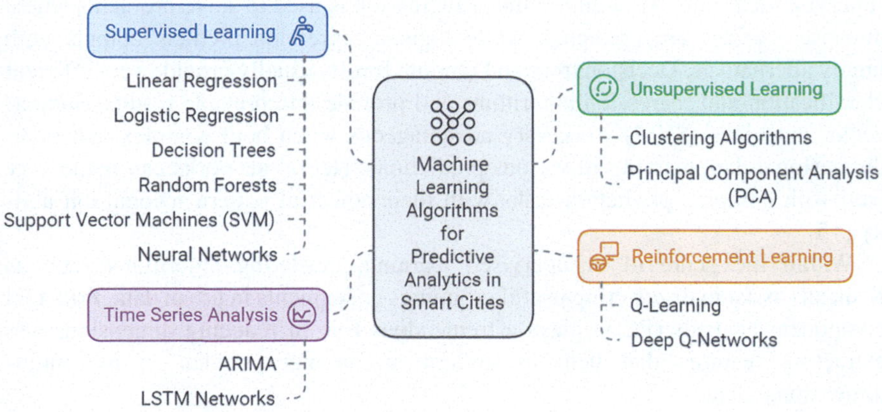

Fig. 4 ML algorithms for predictive analytics in smart cities

executed changing any component would necessarily require changing all the connections that come after it.

An essential advancement has been attained through combining blockchain systems with artificial intelligence in Internet of Things applications. This technological combination has to address multiple challenges such as digital authentication, data verification, automated agreements, distributed processing, data exchange security, and transparency of AI. The new information surge from smart and connected devices demands new data storage solutions; to which the application of blockchain architecture very efficiently responds while ensuring protection concerning both data security and data privacy.

Blockchain can patch this with AI on IoT apps—from medical services to agricultural systems, smart residential technology, and logistics networks [19]. The characteristics—a secure digital authentication, unalterable recording, verification processes, and transparency—find their way to mainstream blockchain applications. The merger of self-executing digital contracts with permanent blockchain records for added security of networks. Still, further to be developed, blockchain technology combines both AI and IoT capabilities, offering augmenting traits that overcome all associated inadequacies of traditional centralized systems.

The main features of blockchain include [20]**:**

- **Immutability**: This property is concerned with preserving the transactional history. As the data is recorded on the blockchain, the chances of changing or deleting it are rare.
- **Transparency**: In scenarios when the blockchain is Public, users can see transaction records. This ensures Openness and accountability in Urban governance.
- **Traceability**: The sequential nature of blockchain, which is timestamped, facilitates auditing and tracking of data changes easily.
- **Consensus Algorithm:** It employs the use of consensus algorithms in which blockchain networks agree and validate the status of data to reduce fraudulent entry.
- **Smart Contract**: Smart contracts are digital protocols and can be further programmed to automatically perform an action based on certain conditions. This property of block chain ensures that operations of cities are seamless and secure.

The following Fig. 5 shows the main properties of blockchain for urban governance.

Integration of blockchain technology promises to better data management, urban planning, and citizen trust efforts [21]. Blockchain offers security for a wide array of data types: citizen identity and personal information, property ownership records, voting systems, environmental monitoring data, and energy consumption details. Other than making that data safer, using a decentralized secure technology would give a lot more credence to the data management of smart cities, which is very crucial for informed decision-making and effective urban operations.

De-centralization increases the resilience of the system, whereby the risk of data loss arising from server failure or attack is greatly reduced. The costs for running de-centralized storage might also drop compared with the cost of having a

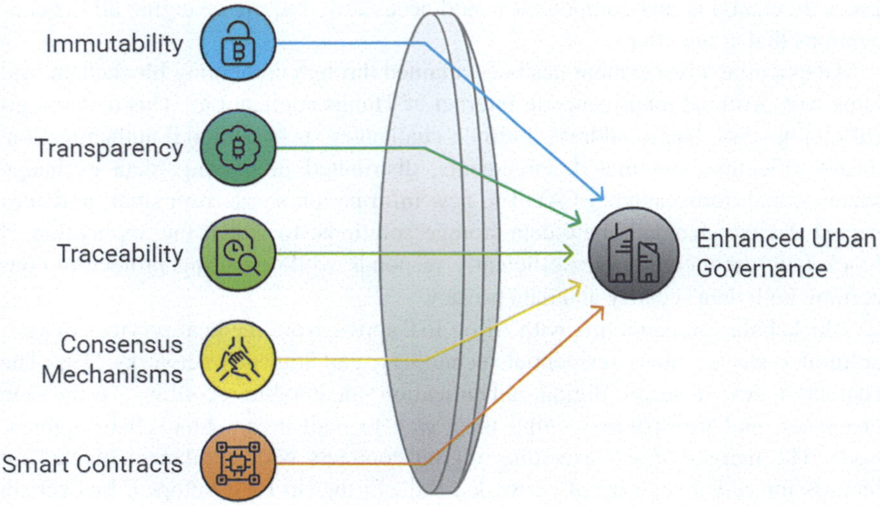

Fig. 5 Blockchain for urban governance in smart cities

traditional central data center, while improving data accessibility and availability. In addition, fragmentation and encryption of the data make it difficult for anyone to manipulate large segments of data in a decentralized framework, and thus, better security. This can lead to shared data within a city easily between departments and other outside parties and promote interoperable with real-time data updates that can serve for urban decision-making purposes [22].

As the expanding field of blockchain technology continues to thrive and evolve, there are gestures toward deeper integration into smart city data management procedures with even more meticulous safe transparent efficient urban governance [23]. Nevertheless, many challenges, including scalability, energy consumption, and law compliance, must be mitigated to ensure the wide application of blockchain technology across urban settings.

2 Smart City Applications

Smart cities are using technologies like AI, ML, and blockchain to cope with diverse urban problems. These advanced solutions aid smart cities in managing critical systems and services and enhancing sustainability, efficiency, and quality of life for citizens.

2.1 *Traffic Management*

AI and ML are transforming traditional approaches to deal with smart cities complex challenges. Intelligent Transportation System uses diverse data sources such as smart card, RFID tags, sensors, video cameras, GPS, and social media to collect required information [24]. Machine learning (ML) allows the IoT sensors to extract meaningful information from an enormous amount of data [25]. Predictive analytics can forecast and optimize traffic signal timing using historical and real-time data [26].

Undoubtedly, intelligent routing suggests personalized routes for drivers in real time, taking consideration of present and predicted traffic conditions. Adaptive traffic signal control employs reinforcement learning algorithms to dynamically adjust signal timings based on the evolving patterns of traffic. Computer vision and machine learning techniques allow for rapid incident detection and response [27], while parking systems redirect drivers toward available free spots, hence restricting congestion caused by search for parking places [28]. Figure 6 below provides a summary about the application of AI in public transport for smart cities, including data sources.

2.2 *Energy Management*

Energy management is key aspect in smart city administration. Load forecasting models predict energy demand to optimize power generation and distribution. Smart grid optimization leverages ML algorithms to balance energy supply and demand in real time, seamlessly integrating renewable energy sources [29]. Building energy management systems implement predictive models to optimize HVAC systems and reduce consumption [30]. Fault detection and predictive maintenance capabilities identify potential failures in energy infrastructure before they occur, minimizing downtime and maintenance costs. Energy Management leverages Neural Networks and advanced anomaly detection systems to optimize power distribution and consumption. These systems analyze smart meter data, weather patterns, and historical usage to predict demand and detect potential faults. Time Series Forecasting helps in grid optimization, ensuring efficient energy distribution while minimizing wastage and maintaining system reliability. Figure 7 presents a summary of data sources and AI algorithms for smart city applications in the energy management domain.

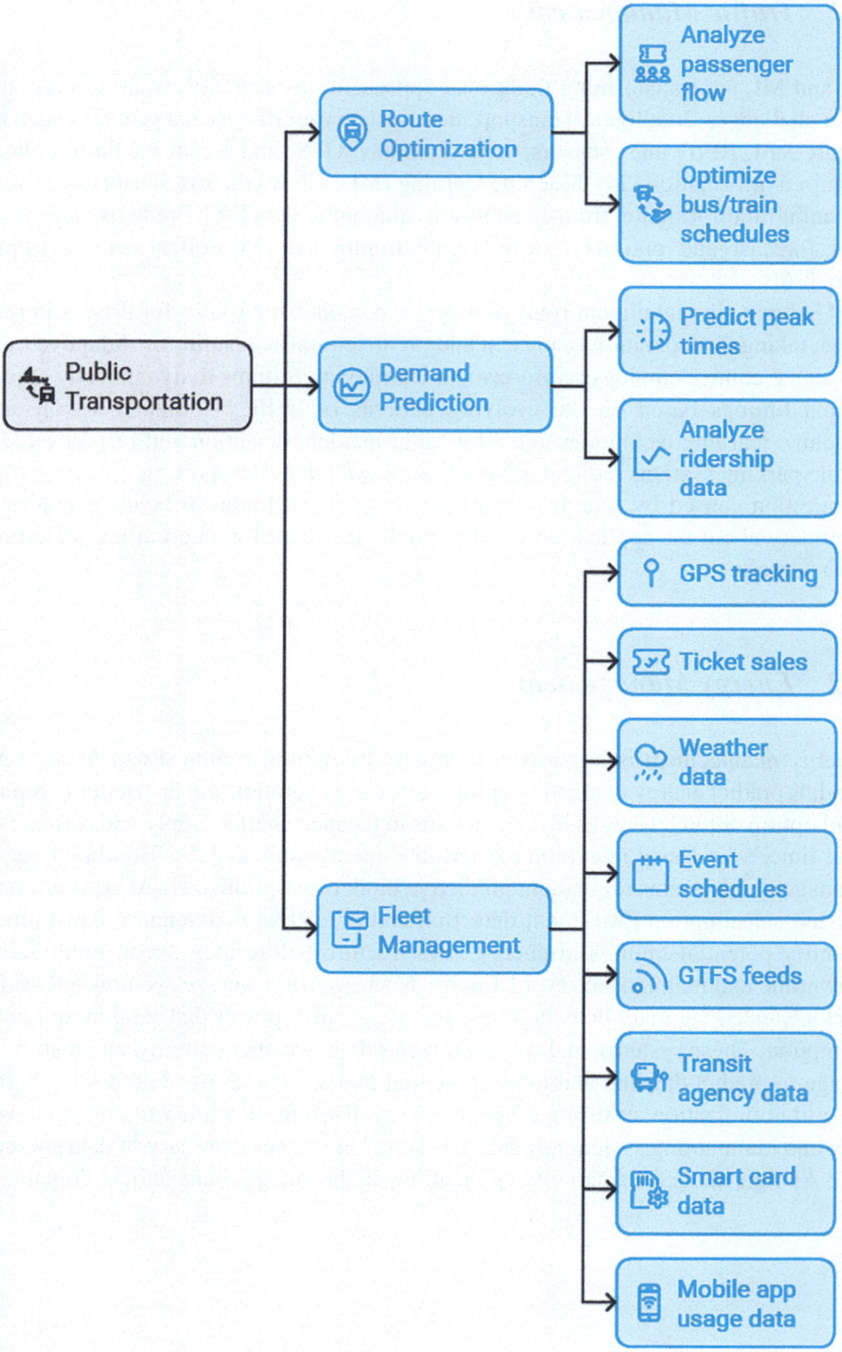

Fig. 6 Application of AI in public transport for smart cities, including data sources

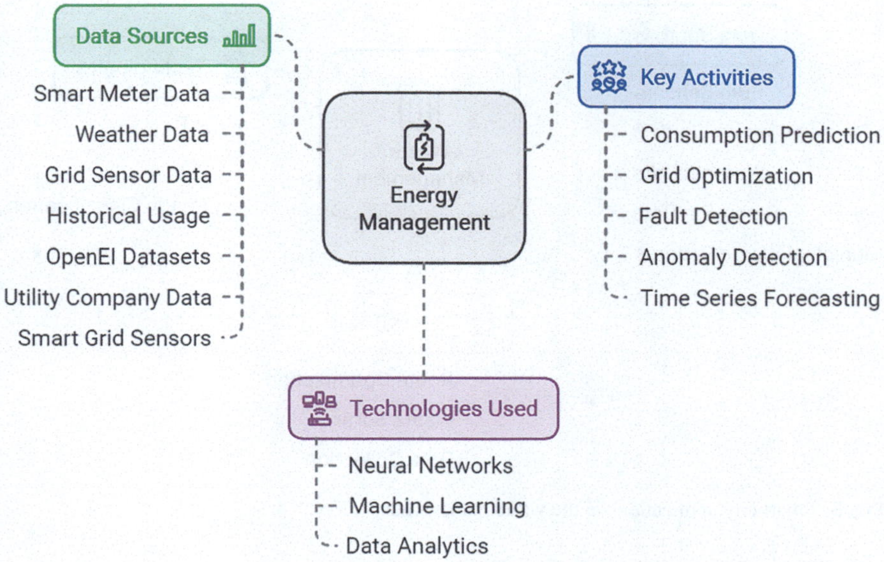

Fig. 7 Application of AI in energy management for smart cities

2.3 *Waste Management*

In waste management domain, AI and ML are enhancing various aspects of the system [31]. Optimized collection routes are determined using predictive analytics based on historical data and real time fill levels of smart bins. Computer vision and ML algorithms automate the sorting of recyclable materials, improving recycling efficiency. Predictive maintenance of the waste management fleet forecasts vehicle maintenance needs, reducing downtime and extending the lifespan of collection vehicles. Waste generation forecasting [32] helps better allocate resources and plan for infrastructure, while smart bin management optimizes collection schedules and reduces overflow incidents. Figure 8 presents a summary of data sources and AI algorithms for smart city applications in the waste management domain.

2.4 *Public Safety*

Public Safety has been enhanced through Predictive Analytics and Computer Vision technologies. These systems process crime records, emergency calls, and social media data to predict potential crime hotspots and manage crowd density. Pattern Recognition algorithms analyze surveillance feeds to detect suspicious activities, enabling faster emergency response and proactive safety measures. Figure 9 presents a summary of data sources and AI algorithms for smart city applications in the public safety domain.

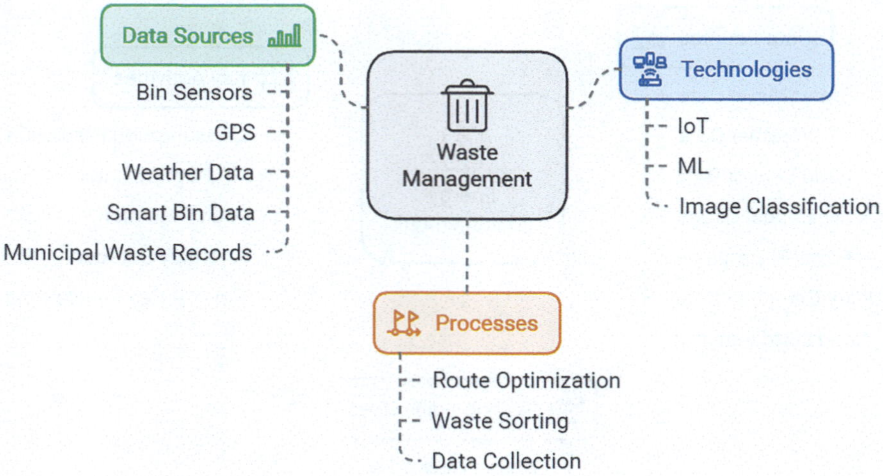

Fig. 8 Smart city applications in the waste management

2.5 *Environmental Monitoring*

Environmental monitoring involves regression models and neural networks to track and predict environmental conditions. Employing sensor fusion technology, information from air quality nonintrusive sensors, weather stations, and noise monitors has been integrated to provide critical insight into the conditions of the environment. This makes it possible for the city to take proactive measures against pollution and guarantee human friendly condition. Figure 10 presents a summary of data sources and AI algorithms for smart city applications in the Environmental Monitoring domain.

2.6 *Building Management*

Building Management has really changed with new age attributed to deep learning and predictive maintenance algorithms. By assessing energy usage patterns, occupancy detection, and equipment performance, it is possible to optimize HVAC operations and predict maintenance needs in buildings. An efficient building management results in improved energy efficiency while making the occupants comfortable and decreasing operation costs. Figure 11 presents a summary of data sources and AI algorithms for smart city applications in the Building Management domain.

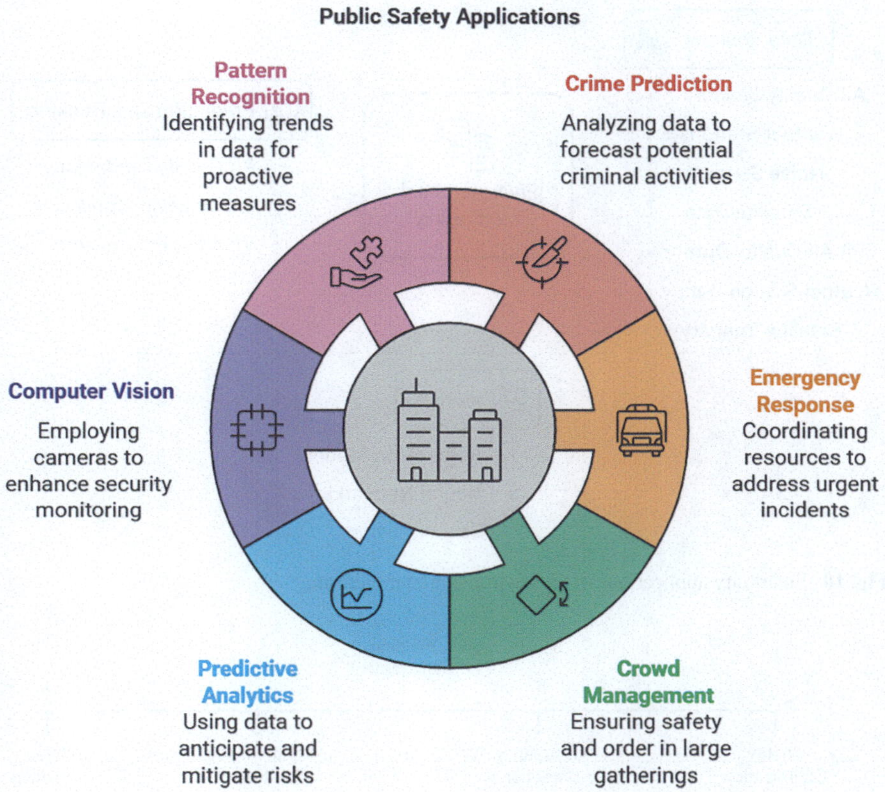

Fig. 9 Smart city applications in the public safety

2.7 *Water Management*

Water Management systems employ: anomaly detection and time series analysis to ensure efficient water distribution and quality. These systems process data from flow sensors, quality monitors, and consumption patterns to detect leaks, predict usage, and maintain water quality standards. Implementing water management approach allows a city to save on water supply issues without compromising supply quality. Figure 12 features a summary of strategies for water management.

2.8 *Citizen Services*

Citizen services have been enhanced through NLP powdered chatbots and sentiment analysis tools. By processing service requests, social media interactions, and feedback forms, these systems will be able to provide 24/7 assistance and predict service needs. Text Classification algorithms help analyze citizen feedback allowing

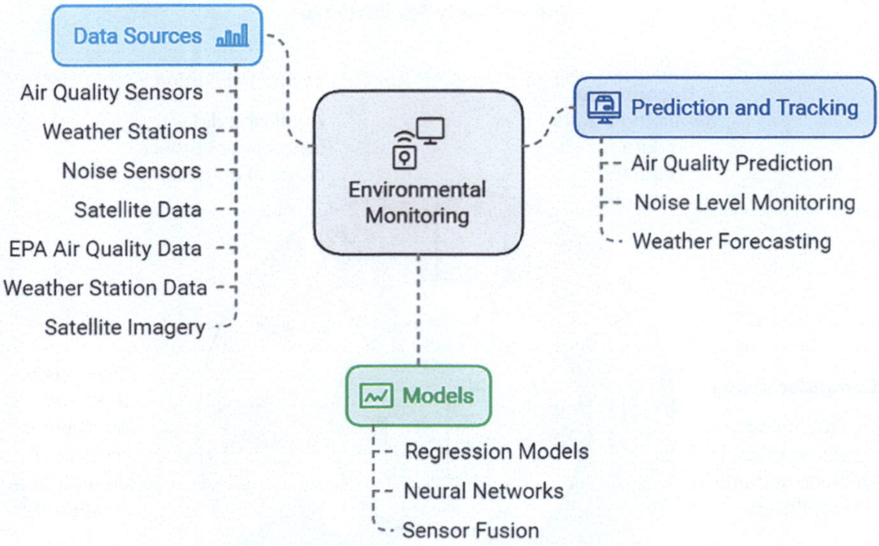

Fig. 10 Smart city applications in the environmental monitoring

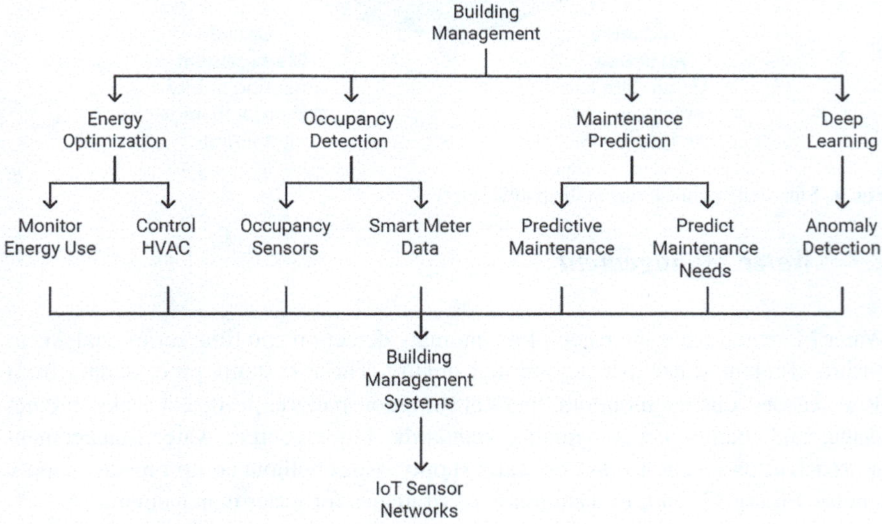

Fig. 11 Smart city applications in the building management

cities to continuously improve delivery of services. Figure 13 provides a summary of the data sources and AI algorithms in smart city applications in the Citizen Services domain.

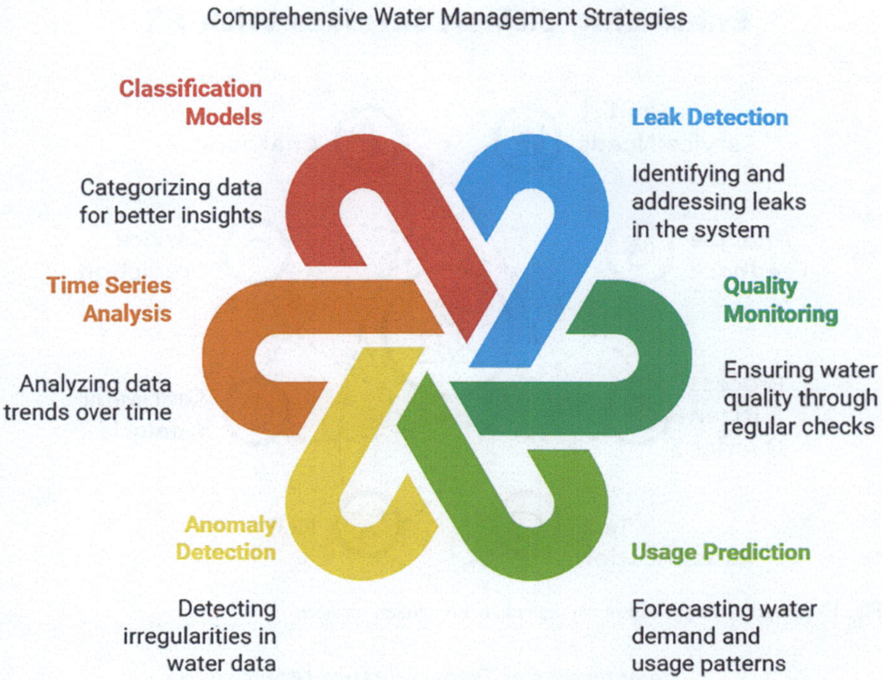

Fig. 12 Water management strategies for smart city applications

2.9 *Infrastructure Monitoring*

Infrastructure monitoring fuses computer vision and predictive analytics to maintain every urban infrastructure. Assorted sensors, inspection findings, and maintenance records are analyzed by these systems to forecast breakdowns or problems, thus schedule preventive maintenance appropriately. This strategy helps a city avoid infrastructure failures and leads to improved resource allocation. Figure 14 provides a summary of data sources and AI algorithms in smart city applications in Infrastructure Monitoring.

2.10 *Emergency Management*

Emergency management utilizes Risk Analysis and Decision Support Systems to perform well on crises impacting cities. The systems analyze weather data, historical events, and population information in order to better predict disasters and plan responses. Resource Optimization algorithms might allocate emergency resources away quickly and effectively from afflicted areas during a crisis. Figure 15 features

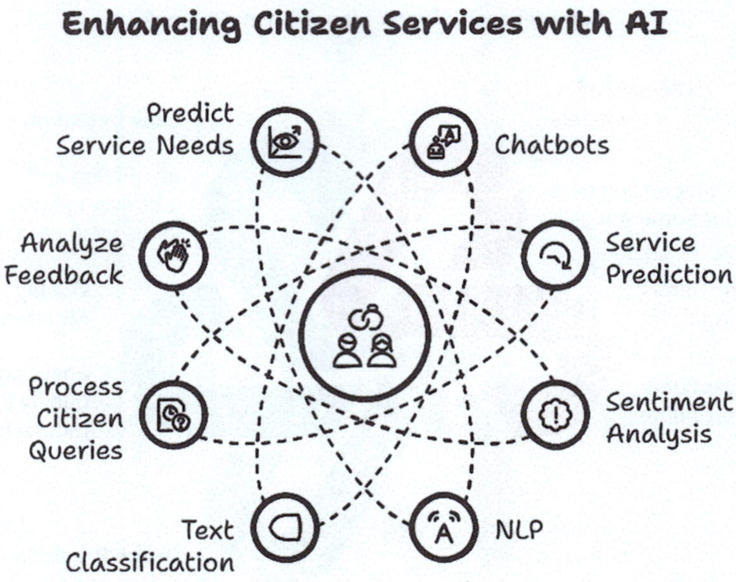

Fig. 13 Smart city applications in the enhancing citizen services

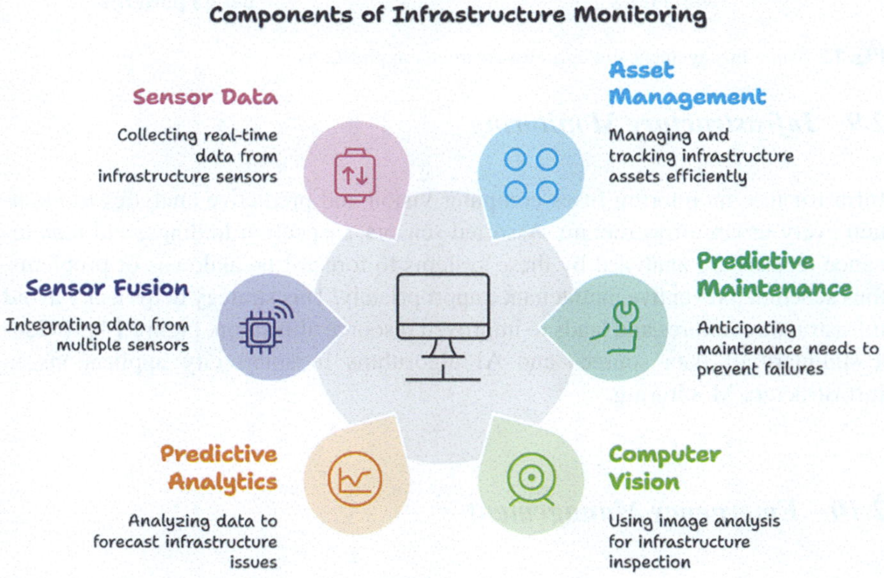

Fig. 14 Smart city applications in infrastructure monitoring

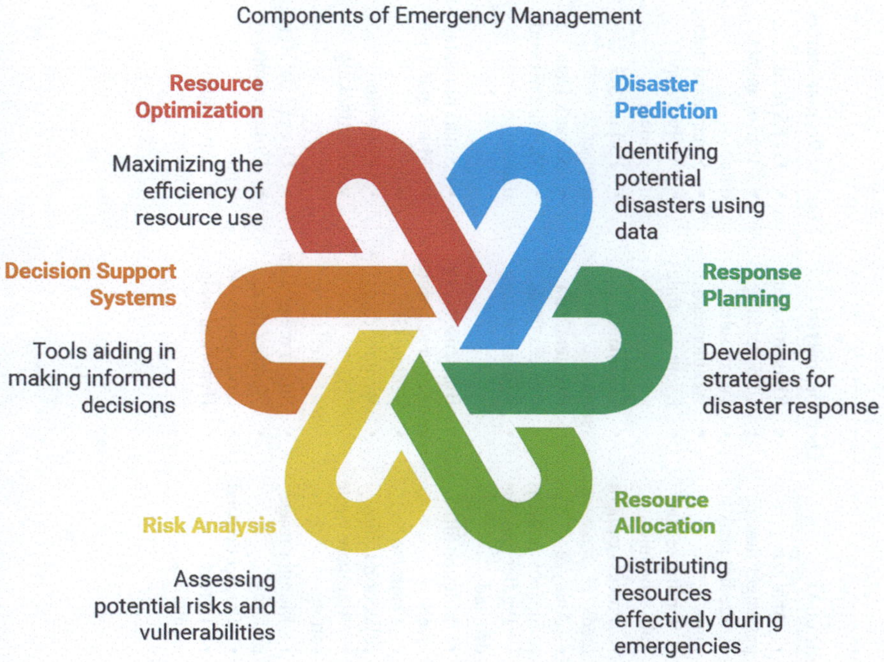

Fig. 15 Smart city applications in emergency management

a summary of the data sources and AI algorithms in smart city applications on Emergency Management.

The following table summarizes AI/ML-Powered Predictive Analytics in Smart City Areas (Table 1).

3 Use Case: Deep Learning Implementation in Traffic Management [47]

The implementation of deep learning in traffic management represents a significant advancement over traditional rule-based systems, offering intelligent, adaptive solutions to modern urban traffic challenges [48, 49]. This framework outlines a systematic approach to developing and deploying AI-powered traffic management systems, encompassing data collection, processing, analysis, and real-time control mechanisms. The following implementation guide and pseudocode provide a foundational structure for building such systems; however, it's crucial to note that this is a high-level framework that requires customization based on specific local requirements, available infrastructure, and operational constraints. Key considerations include hardware capabilities (computing power, sensor networks, storage), data management (quality, privacy, integration), scalability needs, and local traffic patterns.

Table 1 AI-powered predictive analytics in smart city areas

Smart city domain	Sub-domain	AI techniques	Role of AI	Data for analysis	Data collection sources
Public transportation [33, 34]	• Route optimization • Demand prediction • Fleet management	• Random forests • LSTM networks • Gradient boosting	• Analyze passenger flow • Predict peak times • Optimize bus/train schedules	• Ridership data • GPS tracking • Ticket sales • Weather data • Event schedules	• GTFS feeds • Transit agency data • Smart card data • Mobile app usage data
Parking management [35, 36]	• Space availability • Dynamic pricing • Violation detection	• Computer vision • Support vector machines • Neural networks	• Process camera feeds • Analyze occupancy patterns • Predict parking demand	• Occupancy sensors • Payment data • Camera feeds • Historical usage	• ParkingSG dataset • SFpark Data • Parking network data
Waste management [37]	• Collection route optimization • Bin level monitoring • Sorting automation	• IoT + ML • Image classification route optimization algorithms	• Monitor fill levels • Optimize collection routes • Automate waste sorting	• Bin sensor data collection schedules • Vehicle GPS • Weather data	• Smart bin data • Municipal waste records • Collection vehicle tracking
Energy management [38]	• Consumption prediction • Grid optimization • Fault detection	• Neural networks • Anomaly detection • Time series forecasting	• Monitor usage patterns • Predict demand • Detect anomalies	• Smart meter data • Weather data • Grid sensor data • Historical usage	• OpenEI datasets • Utility company data • Smart grid sensors

Smart city domain	Sub-domain	AI techniques	Role of AI	Data for analysis	Data collection sources
Public safety [39]	• Crime prediction • Emergency response • Crowd management	• Predictive analytics • Computer vision • Pattern recognition	• Analyze crime patterns • Monitor crowd density • Detect suspicious activities	• Crime records • Camera feeds • Emergency calls • Social media	• City crime data • Emergency response logs • CCTV footage
Environmental monitoring [40, 41]	• Air quality prediction • Noise level monitoring • Weather forecasting	• Regression models • Neural networks • Sensor fusion	• Monitor pollutant levels • Predict air quality • Track noise pollution	• Air quality sensors • Weather stations • Noise sensors • Satellite data	• EPA air quality Data • Weather station data • Satellite imagery
Building management [42]	• Energy optimization • Occupancy detection • Maintenance prediction	• Deep learning • Anomaly detection • Predictive maintenance	• Monitor energy use • Control HVAC • Predict maintenance needs	• Energy usage • Occupancy sensors • Equipment sensors • Temperature data	• Building management systems • Smart meter data • IoT sensor networks
Water management [43]	• Leak detection • Quality monitoring • Usage prediction	• Anomaly detection • Time series analysis • Classification models	• Monitor flow rates • Detect leaks • Predict demand	• Flow sensors • Quality sensors • Consumption data • Weather data	• Water utility data • Smart meter readings • Quality monitoring

(continued)

Table 1 (continued)

Smart city domain	Sub-domain	AI techniques	Role of AI	Data for analysis	Data collection sources
Citizen services [44]	• Chatbots • Service prediction • Sentiment analysis	• NLP • Text classification • Sentiment analysis	• Process citizen queries • Analyze feedback • Predict service needs	• Service requests • Social media • Feedback forms • Chat logs	• 311 call data • Social media feeds • Citizen feedback
Infrastructure monitoring [45]	• Asset management • Predictive maintenance structural health	• Computer vision • Predictive analytics • Sensor fusion	• Monitor infrastructure predict failures • Track maintenance needs	• Sensor data • Inspection reports • Maintenance logs • Image data	• Infrastructure sensors • Inspection records • IoT device data
Emergency management [46]	• Disaster prediction • Response planning • Resource allocation	• Risk analysis • Decision support systems • Resource optimization	• Predict disasters • Plan responses • Allocate resources	• Weather data • Historical events • Resource inventory • Population data	• Emergency response data • Weather records • Resource databases

Success depends on careful adaptation to local conditions, proper resource allocation, and continuous system optimization. The framework serves as a starting point for municipalities and developers, providing essential components and logical workflows while allowing flexibility for specific implementation needs, regulatory requirements, and technological capabilities. Key steps are depicted in Fig. 16.

3.1 Powered Traffic Flow: Implementation Blueprint and Code Framework

3.1.1 System Initialization and Setup

(a) **Data Collection System Setup:**

- Configure traffic cameras and sensors
- Establish data streaming connections
- Initialize database connections

```
class TrafficDataCollector:
  def __init__(self):
    self.camera_feeds = initialize_cameras()
    self.sensors = initialize_sensors()
```

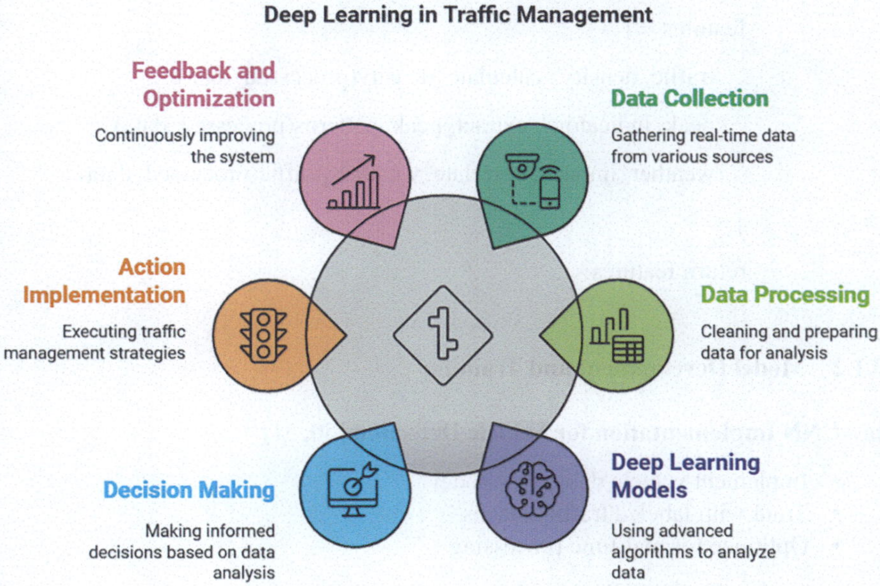

Fig. 16 Steps for implementing deep learning models in traffic management for smart cities

```
self.database = connect_to_database()
    def collect_real_time_data(self):
while True:
    camera_data = self.process_camera_feeds()
    sensor_data = self.get_sensor_readings()
    weather_data = self.get_weather_data()
    self.store_data(camera_data, sensor_data, weather_data)
```

(b) **Data Pipeline Configuration**

- Set up data preprocessing pipelines
- Configure data validation rules
- Establish data transformation workflows

```
class Data Preprocessor:
    def preprocess_data(self, raw_data):
        cleaned_data = remove_noise_and_outliers(raw_data)
        normalized_data = normalize_features(cleaned_data)
        aligned_data = temporal_alignment(normalized_data)
        return aligned_data
    def feature_engineering(self, processed_data):
        features = {
            'traffic_density': calculate_density(processed_data),
            'peak_indicators': extract_peak_patterns(processed_data),
            'weather_impact': correlate_weather_traffic(processed_data)
        }
        return features
```

3.1.2 Model Development and Training

(a) **CNN Implementation for Vehicle Detection** [50, 51]

- Implement vehicle detection model
- Train with labeled traffic images
- Optimize for real-time processing

```
class TrafficCNN:
    def __init__(self):
```

```
        self.model = build_cnn_architecture()
    def train_model(self, training_data, labels):
        self.model.compile(optimizer='adam',
                           loss='categorical_crossentropy')
        self.model.fit(training_data, labels,
                       epochs=100,
                       validation_split=0.2)
    def detect_vehicles(self, frame):
        preprocessed_frame = preprocess_image(frame)
        predictions = self.model.predict(preprocessed_frame)
        return process_detections(predictions)
```

(b) **LSTM Implementation for Traffic Pattern Analysis** [52]

- Develop sequence prediction model
- Train with historical traffic data
- Implement pattern recognition

```
class TrafficLSTM:
    def __init__(self):
        self.model = build_lstm_architecture()
    def train_pattern_recognition(self, sequence_data):
        self.model.compile(optimizer='rmsprop',
                           loss='mse')
        self.model.fit(sequence_data['X'],
                       sequence_data['y'],
                       batch_size=32,
                       epochs=50)
    def predict_traffic_patterns(self, current_sequence):
        return self.model.predict(current_sequence)
```

3.1.3 Real-Time Processing System

(a) **Traffic Flow Analysis.**

- Implement real-time traffic monitoring

- Calculate traffic metrics
- Detect anomalies

```
class TrafficAnalyzer:
    def analyze_traffic_flow(self, current_data):
        flow_metrics = {
            'density': calculate_traffic_density(current_data),
            'speed': calculate_average_speed(current_data),
            'congestion': detect_congestion(current_data)
        }
        return flow_metrics
    def detect_anomalies(self, flow_metrics):
        anomalies = []
        for metric, value in flow_metrics.items():
            if is_anomalous(value):
                anomalies.append((metric, value))
        return anomalies
```

(b) **Signal Control Optimization**

- Implement adaptive signal timing
- Optimize intersection performance
- Handle emergency priorities

```
class SignalController:
    def optimize_signals(self, traffic_data, predictions):
        for intersection in self.managed_intersections:
            current_flow = analyze_intersection_flow(intersection)
            predicted_flow = predictions[intersection.id]
            optimal_timing = calculate_optimal_timing(
                current_flow,
                predicted_flow
            )
            intersection.update_timing(optimal_timing)
```

```
def handle_emergency_priority(self, emergency_vehicle):
    route = emergency_vehicle.route
    for intersection in route:
        intersection.set_emergency_priority()
```

3.1.4 Decision Support and Action Implementation

(a) **Traffic Management Decisions**

- Implement decision-making logic
- Generate traffic advisories
- Coordinate response actions

```
class TrafficManager:
    def make_decisions(self, analysis_results):
        decisions = {
            'signal_changes': [],
            'route_advisories': [],
            'emergency_responses': []
        }
        if analysis_results['congestion_level'] > THRESHOLD:
            decisions['signal_changes'].append(
                generate_signal_adjustment(analysis_results)
            )
            decisions['route_advisories'].append(
                generate_alternative_routes(analysis_results)
            )
        return decisions
```

(b) **Communication and Control**

- Implement control actions
- Update message signs
- Send notifications

```
class CommunicationController:
    def update_message_signs(self, advisories):
        for sign in self.message_signs:
            relevant_message = generate_message(
                advisories,
                sign.location
            )
            sign.display_message(relevant_message)

    def send_notifications(self, alerts):
        for alert in alerts:
            if alert.priority == 'high':
                send_emergency_notification(alert)
            else:
                send_standard_notification(alert)
```

3.1.5 System Monitoring and Optimization

(a) **Performance Monitoring**

- Track system performance
- Collect metrics
- Generate reports

```
class PerformanceMonitor:
  def collect_metrics(self):
    metrics = {
      'response_time': measure_response_time(),
      'prediction_accuracy': evaluate_predictions(),
      'system_efficiency': calculate_efficiency()
    }
    self.store_metrics(metrics)
    return metrics
```

(b) **Continuous Learning**

- Update models with new data
- Optimize parameters
- Improve accuracy

```
class ModelOptimizer:
  def update_models(self, performance_metrics):
```

```
        if performance_metrics['accuracy'] < THRESHOLD:
            new_training_data = collect_recent_data()
            retrain_models(new_training_data)
    def optimize_parameters(self):
        current_performance = evaluate_system_performance()
        optimal_params = find_optimal_parameters(current_performance)
        update_system_parameters(optimal_params)
```

This implementation guide provides a structured approach to developing a deep learning-based traffic management system, with each component designed to work together in a cohesive system. The pseudocode demonstrates the key algorithms and processes needed for implementation.

4 Case Studies

4.1 Smart City Implementation in Lusail: A Case Study of AI-Driven Urban Transformation

This case study traces the implementation of an artificial intelligence-driven smart city platform in Lusail City, Qatar, which considers the change in urban digital combinatorialities. It further examines ST Engineering's $60 million project of building, designing, and running the smart city infrastructures in whole, addressing the integration of these existing smart technologies like AI, machine learning, and data analytics formulating urban cities [53, 54].

Urban digitalization has become integral to modern city construction and smart city initiatives as base drivers of sustainable urbanization. In this sense, the city considered one of the biggest sustainable urban developments in Qatar, Lusail City, launched its transformation program to create an integrated smart city in October 2024. This is an important initiative within Qatar's National Vision 2030, providing insight into the application of advanced technologies in the urban context.

Lusail City spans an area of 38 sq km, containing four unremittingly conventional islands and 19 multifunctional districts for residential, commercial, and entertaining use. The smart city program, under the big umbrella of ST Engineering Urban Solutions, focuses, among other things, on AGIL Smart City Operating System (OS), to serve as the digital backbone of the operation of this city. The project will, therefore, run from Q4 2024 to 2027 and subsequently seeks to improve the lives of about 450,000 residents and visitors while promoting the goals of sustainable urbanization.

The core of the project encompasses an AGIL Smart city Operating System, AGIL, with all interconnected complex urban systems linked into the OS. Lighting,

BMS, traffic control-this is integrated within the platform together with the capability to connect to an Asset Management System, ensuring continuous automated monitoring and protection of critical infrastructures. It adopts advanced workflow automation backed with AI and provides city-level insight into maximizing resource utilization and enhancing operational efficiency.

What distinguishes this implementation is the introduction of resident-and-visitor-centric interactive applications that include real-time feedback systems and AI-enabled, natural language communication chatbots.

4.2 *Satellite-Enabled Smart City Management in Bahrain*

Modern urban management increasingly relies on advanced technological solutions to monitor and guide development. In 2024, the Kingdom of Bahrain implemented a sophisticated urban planning system utilizing satellite data and artificial intelligence, representing a significant advancement in smart city management. This implementation offers valuable insights into the application of space-based technologies for urban monitoring and development control [55].

The initiative centers on the integration of Planet's SkySat 50 cm high-resolution satellite imagery with Aetosky's AI-enabled change detection technology. The system, implemented by the Survey and Land Registration Bureau (SLRB), provides comprehensive coverage of Bahrain's urban areas and enables automated monitoring of various urban changes, including building modifications, rooftop alterations, vegetation changes, and water related developments.

The core of the project lies in the integration of multiple technological components:

- **Satellite Data Infrastructure**: The system utilizes SkySat's 50 cm resolution imagery, providing detailed visual data of urban areas.
- **Data Management System**: A specialized data storage and organization system allows for the cataloguing and management of spatial and temporal datasets, enabling comprehensive nationwide analysis.
- **AI-Powered Insights**: Sophisticated change detection algorithms automatically identify and classify diverse urban transformations, ranging from construction projects to environmental shifts.
- **Consistent Monitoring**: Biweekly data updates facilitate the continuous tracking and analysis of developments, infrastructure modifications, and environmental alterations.

The implementation has already demonstrated significant benefits in urban management efficiency:

- **Enhanced Building Permit Validation**: Municipalities have reported increased effectiveness in validating building permits through automated change detection.

- **Resource Optimization**: The system has reduced monitoring time and enabled more cost-effective resource allocation.
- **Environmental Monitoring**: Improved tracking of urban greenspaces and coastline activities supports environmental protection initiatives.
- **Cross-Departmental Collaboration**: The SLRB assists as a centralized data hub supporting various government departments in environmental monitoring, transportation, and municipal response.

Figure 17 shows integrated satellite imagery with AI technology.

The project's success is supported by the collaboration of established organizations:

- Planet: A leading provider of satellite imagery and data.
- Aetosky (formerly SkyMap Global): Specialists in AI-enabled change detection technology.
- SLRB: Bahrain's authoritative body for land registration and survey.

The implementation of satellite-based urban monitoring in Bahrain marks a significant leap forward in smart city management. By combining high-resolution satellite imagery with AI-powered change detection, this system provides a powerful tool for urban planning and monitoring. This advanced technology offers numerous benefits, including increased efficiency, improved accuracy, and enhanced decision-making capabilities. This innovative approach not only supports the Kingdom's sustainable development goals but also provides valuable insights for other regions considering similar technological advancements. Future research should focus on quantifying the long-term impacts of this technology on urban development, environmental protection, and resource optimization.

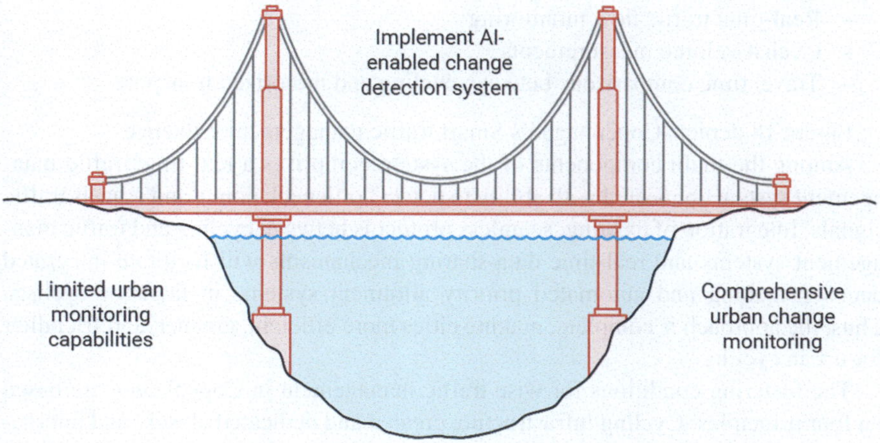

Fig. 17 Integrated satellite imagery with AI technology

4.3 Copenhagen's Smart Traffic Management: Climate-Driven Urban Mobility

Urban mobility plays a crucial role in achieving climate objectives. Copenhagen's commitment to becoming carbon neutral by 2025 has led to innovative approaches in traffic management and cycling infrastructure. This implementation offers valuable insights into how cities can leverage smart technology to promote sustainable transportation while improving urban mobility [56].

The initiative centers on two main components: data-driven traffic management through the MobiMaestro platform and enhanced cycling infrastructure. The project, implemented through a long-term partnership between Copenhagen and Technolution Move since 2015, aims to optimize traffic flow while promoting sustainable transportation methods, particularly cycling.

The system integrates several key technological components:

- **MobiMaestro Platform**
  - Centralized traffic management system
  - Real-time monitoring capabilities
  - Integration of Floating Car Data (GPS-based vehicle tracking)
  - Automated and manual traffic regulation adjustment
- **Cycling Infrastructure**
  - Dynamic Variable Message Signs (VMS) specifically designed for cyclists
  - BikeSim simulator integration
  - Real-time traffic light controllers
  - Sensor network for cyclist monitoring
- **Data Collection and Analysis**
  - GPS-based vehicle tracking
  - Real-time traffic flow monitoring
  - Cyclist volume measurements
  - Travel time comparisons between cycling and motorized transport

Figure 18 depicts Copenhagen's Smart traffic management sequence.

Among the main components of the system comprises a real-time traffic management system connected with ability to detect bottleneck points and adjusts traffic signals, Integration of existing, seamless protocols between cycling and traffic management systems and real-time data-sharing mechanisms will facilitate integrated decision-making and automated priority allotment systems in favor of cyclists. Thus, the approach is complete, making cities more efficient, greener, and friendlier for urban cycling.

The fostering conditions for wise traffic management in Copenhagen are based on four principles: Cycling infrastructure created and dedicated closely and internationally recognized as such; integrated traffic management systems; signal

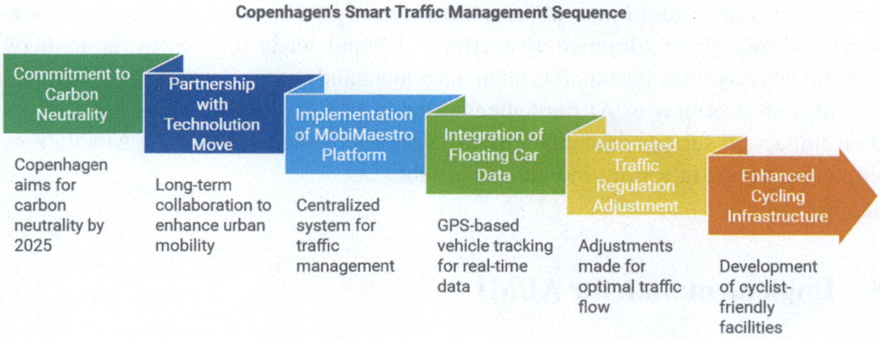

Fig. 18 Smart traffic management: Copenhagen's city

optimization in real time; and data-driven decision-making for promoting more sustainable-choice transport modes.

The final aspect of what Copenhagen's success achieves is allowing the smart traffic solutions to dovetail nicely with environmental objectives. It is a combination of the data-driven framework for traffic management with support for cycling infrastructure to provide an overall communication concept that links the objectives of urban mobility with carbon neutrality. This case set up a good example and worthy lesson for other cities looking at implementing sustainable transport solutions.

4.4 AI Implementation in Indian Smart Cities: A Comprehensive Analysis

This case study considers the implementation of AI solutions across various Indian smart cities, focusing on key urban domains including mobility, safety and security, waste management, education, e-governance, and energy management.

India's smart cities are adopting AI-driven solutions in multiple domains, demonstrating a wide variety of applications and implementations around the country. In the case of mobility, twelve cities, including major metro areas such as Bengaluru, Ahmedabad and Varanasi, are implementing AI systems to support traffic flow, public transport and urban mobility infrastructure. Such implementations feature the range from intelligent traffic management systems to smart parking solutions.

Safety measures, from security assurances, have received top priority, in five cities including Varanasi and Visakhapatnam, which have integrated AI-driven surveillance and emergency response systems to enhance citizen safety. Innovative implementations of AI applications have been made in waste management in Varanasi and Visakhapatnam, through real-time collection routing optimizing and intelligent monitoring systems. In the education sector alone, Belagavi and Pimpri Chinchwad have undertaken implementing AI supported learning solutions and smart classroom initiatives. E-Governance has come big in tech hubs such as

Bengaluru, Jaipur, and Pune, utilizing AI-driven systems to improve citizens' services and minimize administrative efforts. Bhopal leads the energy domain by adopting AI solutions for Smart Grid management and energy optimization. Figure 2 provides an overview of AI implementations in Indian cities by key solutions and their impact on various domains. Figure 19 shows the domain specific AI solutions implementations in several Indian cities (Table 2).

5 Implementation for AI/ML

(a) **Technology Implementation**: In predictive analytics, algorithms are used to know what is likely to occur, thus enabling automated decisions and improving the efficiency of urban services operations. There are various layers of infrastructure, which form the base of data collection and monitoring. There would be visual streams through CCTV cameras, environmental sensory data, and

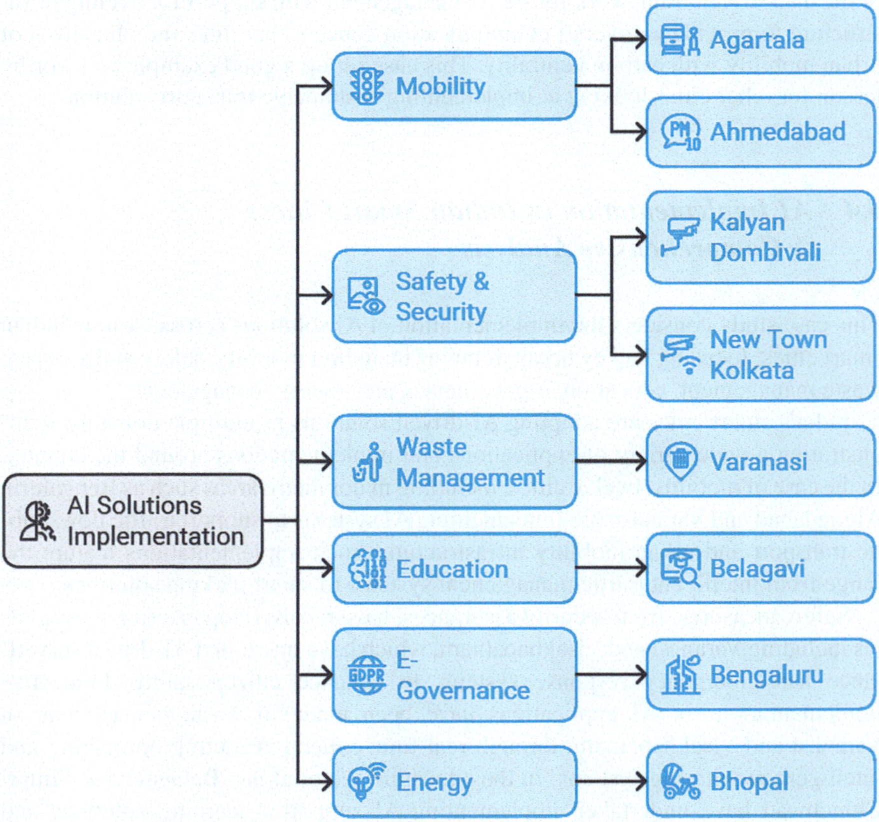

Fig. 19 Implementation of AI in smart cities in India

Table 2 Comprehensive overview of AI implementations across smart cities in India

Domain	Cities	Key solutions	Impact
Mobility	Agartala, Ahmedabad, Bengaluru, Jabalpur, Nagpur, Pimpri Chinchwad, Ranchi, Satna, Surat, Tumakuru, Varanasi, Visakhapatnam	ATCS (adaptive traffic control system) ANPR (automatic number plate recognition) RLVD (red light violation detection) AI-enabled traffic monitoring	Reduced wait times at signals (5070%) Improved traffic flow Better violation detection Enhanced traffic rule enforcement
Safety and security	Kalyan Dombivali, New Town Kolkata, Pimpri Chinchwad, Varanasi, Visakhapatnam	AI-powered surveillance Mask detection systems Crowd monitoring Intrusion detection	Enhanced public safety Better pandemic management Improved emergency response Reduced crime rates
Waste management	Varanasi, Visakhapatnam	Smart bin monitoring Waste collection optimization AI-based cleanliness monitoring	Improved waste collection efficiency Better resource allocation Cleaner city spaces
Education	Belagavi, Pimpri Chinchwad	AI-enabled e-learning Student behavior analysis Teaching effectiveness monitoring	Improved learning outcomes Better student engagement Enhanced teaching quality
E-governance	Bengaluru, Jaipur, Pune	AI chatbots Sentiment analysis Property tax assessment Attendance management	Better citizen service delivery Improved revenue collection Enhanced administrative efficiency
Energy	Bhopal	AI-powered smart poles Smart lighting Environmental monitoring	46% annual energy savings Enhanced city surveillance Improved environmental monitoring

operational utilization metrics-gathered through IoT sensors. Smart poles are multi-utility units that will serve different services, while control centers tie everything together and act as a nervous system at a city level.

An integration mechanism will link the services by ICCC and ensure connectivity of all departments under one roof and seamless coordination to achieve central monitoring. This brings about a common layer of operation across all departments of the city, sustaining the existing state for better coordination between them and speedily solving urban woes. Truly, this interconnectedness is what establishes a sound ecosystem, enabling seamless flow of data for real-time decision-making and proactive urban management. Figure 20 shows AI/ML System and Infrastructure Integration.

(b) **Implementation Challenges:** Smart city implementations face three primary categories of challenges: technical, operational, and infrastructural. Technical challenges involve integrating new AI systems with legacy infrastructure and maintaining data quality, addressed through phased implementations and regular updates. Operational challenges focus on user adoption and training needs, mitigated through comprehensive stakeholder engagement and structured train-

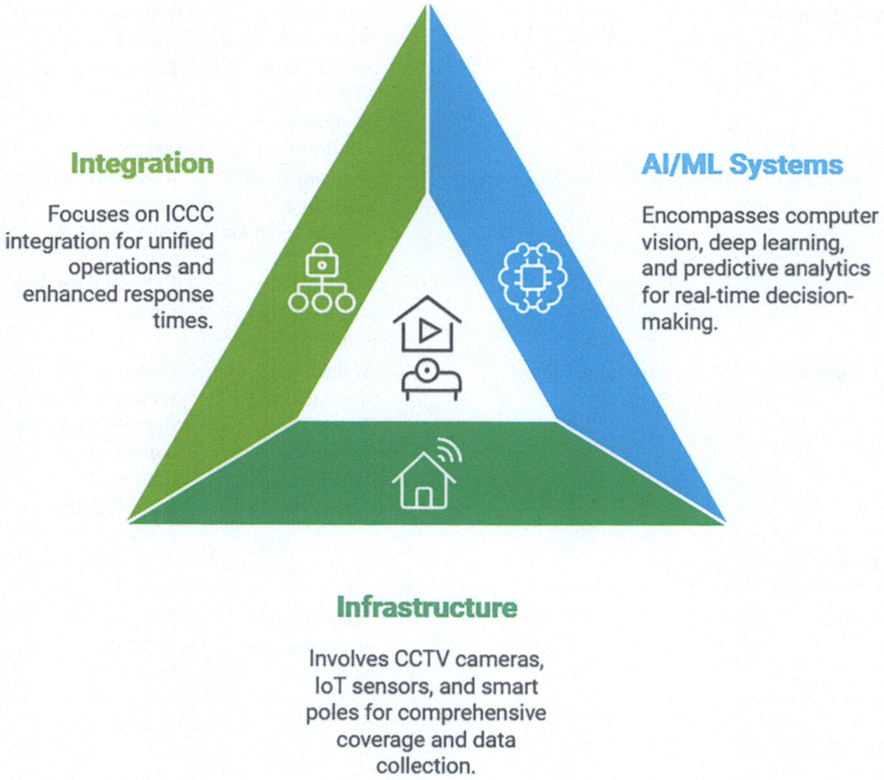

Fig. 20 AI/ML system and infrastructure integration

ing programs. Infrastructure challenges concern physical implementation aspects like hardware installation and network coverage, resolved through careful planning and robust maintenance protocols. Success relies on addressing these challenges through systematic mitigation strategies while ensuring sustainable and efficient urban operations (Table 3).

6 Discussions

Support by leadership is one of the hallmarks of the successful smart city initiative. When the city government strongly supports the implementation, it eases the path of implementation and allows better allocation of all required resources. When the leaders of the city stand in support of these AI-based initiatives, funds for the implementation of different projects are disbursed in a proper way, bureaucratic obstructions are markedly lowered, and approvals are provided very swiftly. This support does provide with a conducive environment for innovation development and cross-departmental collaboration.

Technology selection serves as a key player in the accomplishment of desired results. Cities need to prospect potential AI solutions so that they will serve their very distinct challenges and capabilities of the underlying infrastructure. Proper technology formalization will ensure an optimal return on investment, owing to the somewhat costly mismatches of a solution and needs being avoided. This implies scrupulous consideration of the various AI technologies that are possible to consider. There is also the consideration of scalability and long-term maintenance requirements.

Stakeholder engagement across different departments initiate planning and implementation ensuring a coordinated effort and greater success for AI solutions. Collaboration helps in spotting problems early, gathering all requirements, and helps integrate new systems back into the process.

Citizen engagement turns out to be the vital for smart city success. All processes pertaining to planning, consultation, and feedback with proper inclusion of citizens provide the structure for ensuring that the implemented AI actually serves the needs

Table 3 Smart cities implementation challenges

Challenge category	Specific issues	Mitigation strategies
Technical	Integration with legacy systems Data quality and accuracy Network connectivity	Phased implementation Regular system updates Robust testing
Operational	User adoption Training requirements Process changes	Stakeholder engagement Comprehensive training Change management
Infrastructure	Hardware installation Network coverage Power supply	Proper planning Backup systems Regular maintenance

of the community. With this level of citizen participation, the likelihood of actually using these smart services with proper tracking and feedback for improvements increases. This facilitates a feedback loop whereby services will actually tailor their offerings based on real user needs and experiences.

7 Conclusion

The paper illustrates how AI along with predictive analytics impact transformation in urban governance decision-making processes even in more areas of public sector activity. The examples provided indicate how these applications are improvements that transform, rather than supplemented, traffic, public safety, waste management, and energy efficiency operations in a city. Data-driven insights are sources of tools for municipal authorities that empower them to allocate their resources more accurately and engage citizens and enable best sustainable practices in cities. The study points to three key paths for future development: standards-based strategies and frameworks, systematic scaling, and continuous technological advances. All of which need to go along an international paradigm that mandates interoperability, adequate security, and privacy protection. Successful scale-up of AI implementations should take into consideration local contexts and requirements such that solutions would be adaptable and sustainable across a multitude of urban contexts. Innovation focused on quantum computing, advanced IoT sensors, and Edge computing would be critical for moving to predictive models with higher sophistication and real-time processing capabilities. However, there are challenges that must be given some thought for smart cities. The most critical of these challenges are data privacy issues system integration complexities and ethics in the use of AI. Future research will explore addressing these issues while addressing cross-domain solutions, which can simultaneously penetrate multiple urban challenges. At the same time, robust governance frameworks will be developed to ensure AI-based transformations in cities are ethical and integrate citizen rights.

References

1. Siman, E., Abiodun, J., Timothy, G., & Nandom, S. (2023). IoT-driven smart cities: Enhancing urban sustainability and quality of life.
2. Herath, K., & Mittal, M. (2022). Adoption of artificial intelligence in smart cities: A comprehensive review. *International Journal of Information Management Data Insights, 2*, 1–21. https://doi.org/10.1016/j.jjimei.2022.100076
3. Bibri, S. E. (2021). Data-driven smart sustainable cities of the future: Urban computing and intelligence for strategic, short-term, and joined-up planning. *Computational Urban Science, 1*, 8. https://doi.org/10.1007/s43762-021-00008-9

4. Potter, K., & Broklyn, P. (2024). AI-based predictive maintenance in manufacturing industries background: AI-based predictive maintenance in Manufacturing Industries. *Artificial Intelligence.*
5. Shankaran R, S., & Rajendran, L. (2021). Real-time adaptive traffic control system for smart cities. Presented at the 2021 International Conference on Computer Communication and Informatics (ICCCI), pp. 1–6. https://doi.org/10.1109/ICCCI50826.2021.9402597.
6. Idrees, S. M., Alam, M. A., Agarwal, P., & Ansari, L. (2019). Effective predictive analytics and modeling based on historical data. In *Advances in computing and Data sciences: Third international conference, ICACDS 2019, Ghaziabad, India, April 12–13, 2019, revised selected papers, part II 3* (pp. 552–564). Springer.
7. Štrumbelj, S. Ž. E. (n.d.). Predictive modelling | Introduction to data science.
8. paula. (2023). *Machine learning models for precise predictive analytics* [WWW Document]. Stefanini. Retrieved from Nov 2, 2024, from https://stefanini.com/en/insights/news/machine-learning-models-for-precise-predictive-analytics
9. Davies, C. (2024). Big Data analytics for smart cities. *International Journal of Computing and Engineering, 6*, 14–29. https://doi.org/10.47941/ijce.2057
10. Casali, Y., Aydin, N. Y., & Comes, T. (2022). Machine learning for spatial analyses in urban areas: A scoping review. *Sustainable Cities and Society, 85*, 104050. https://doi.org/10.1016/j.scs.2022.104050
11. Dinh, T., & Thai, M. (2018). AI and Blockchain: A disruptive integration. *Computer, 51*, 48–53. https://doi.org/10.1109/MC.2018.3620971
12. Bhatia, V., & Jaglan, V. (n.d.). Integrating artificial intelligence and IoT for smart cities.
13. Dou, X., Chen, W., Zhu, L., Bai, Y., Li, Y., & Wu, X. (2023). Machine learning for smart cities: A comprehensive review of applications and opportunities. *International Journal of Advanced Computer Science and Applications, 14.* https://doi.org/10.14569/IJACSA.2023.01409104
14. Theng, M., & Theng, D. (2020). Machine learning algorithms for predictive analytics: A review and new perspectives. https://doi.org/10.37896/HTL26.06/1159.
15. Band, S. S., Ardabili, S., Sookhak, M., Chronopoulos, A. T., Elnaffar, S., Moslehpour, M., Csaba, M., Torok, B., Pai, H.-T., & Mosavi, A. (2022). When smart cities get smarter via machine learning: An in-depth literature review. *IEEE Access, 10*, 60985–61015. https://doi.org/10.1109/ACCESS.2022.3181718
16. Kulkarni, G., Muley, A., Deshmukh, N., & Bhalchandra, P. (2018). Autoregressive integrated moving average time series model for forecasting air pollution in Nanded city, Maharashtra, India. *Modeling Earth Systems and Environment, 4.* https://doi.org/10.1007/s40808-018-0493-2
17. Lu, S., Zhang, Q., Chen, G., & Seng, D. (2021). A combined method for short-term traffic flow prediction based on recurrent neural network. *Alexandria Engineering Journal, 60*, 87–94. https://doi.org/10.1016/j.aej.2020.06.008
18. Satybaldy, A., Subedi, A., & Idrees, S. M. (2024). Decentralized key management for digital identity wallets. In *Blockchain transformations: Navigating the decentralized protocols era* (pp. 47–58). Cham.
19. Salama, R., Altorgoman, A., & Al-Turjman, F. (2024). Chapter 24—An overview of how AI, Blockchain, and IoT are making smart healthcare possible. In F. Al-Turjman (Ed.), *Computational intelligence and Blockchain in complex systems, advanced studies in complex systems* (pp. 255–267). Morgan Kaufmann. https://doi.org/10.1016/B978-0-443-13268-1.00016-9
20. Treiblmaier, H., & Sillaber, C. (2021). The impact of blockchain on e-commerce: A framework for salient research topics. *Electronic Commerce Research and Applications, 48*, 101054. https://doi.org/10.1016/j.elerap.2021.101054
21. Zhu, X., Chen, Z., Cheng, T., Yang, C., Wu, D., Wu, Y., & Wang, H. (2024). Blockchain for urban governance: Enhancing trust in smart city systems with advanced techniques. *Sustainable Cities and Society, 108*, 105438. https://doi.org/10.1016/j.scs.2024.105438

22. Alnahari, M. S., & Ariaratnam, S. T. (2022). The application of Blockchain technology to Smart City infrastructure. *Smart Cities, 5*, 979–993. https://doi.org/10.3390/smartcities5030049
23. DBT, D. (2023). *How smart cities use Blockchain*. Deltec Bank and Trust. Retrieved from Feb 2, 2024, from https://www.deltecbank.com/news-and-insights/how-smart-cities-use-blockchain/
24. Bhatia, V., Jaglan, V., Kumawat, S., Siwach, V., & Sehrawat, H. (2022). Intellıgent transportation system applications: A traffic management perspective. pp. 419–433. https://doi.org/10.1007/978-981-16-2422-3_33.
25. Bhatia, V., & Bhatia, B. (2024). Machine learning-based solutions for internet of things-based applications. In A. K. Tyagi (Ed.), *Automated secure computing for next-generation systems* (pp. 295–318). Wiley. https://doi.org/10.1002/9781394213948.ch15
26. Almukhalfi, H., Noor, A., & Noor, T. H. (2024). Traffic management approaches using machine learning and deep learning techniques: A survey. *Engineering Applications of Artificial Intelligence, 133*, 108147. https://doi.org/10.1016/j.engappai.2024.108147
27. Yijing, H., Wei, W., He, Y., Qihong, W., & Kaiming, X. (2023). Intelligent algorithms for incident detection and management in smart transportation systems. *Computers and Electrical Engineering, 110*, 108839. https://doi.org/10.1016/j.compeleceng.2023.108839
28. Parashar, S., Kumar, G. (2019). Smart parking system using genetic optimization: A review. *Presented at the 2019 International Conference on Intelligent Sustainable Systems (ICISS)*, pp. 599–603. https://doi.org/10.1109/ISS1.2019.8908120.
29. Jain, A., & Gupta, S. C. (2024). Evaluation of electrical load demand forecasting using various machine learning algorithms. *Frontiers in Energy Research, 12*. https://doi.org/10.3389/fenrg.2024.1408119
30. Mariano-Hernández, D., Hernández-Callejo, L., Zorita-Lamadrid, A., Duque-Pérez, O., & Santos García, F. (2021). A review of strategies for building energy management system: Model predictive control, demand side management, optimization, and fault detect & diagnosis. *Journal of Building Engineering, 33*, 101692. https://doi.org/10.1016/j.jobe.2020.101692
31. Alsabt, R., Alkhaldi, W., Adenle, Y. A., & Alshuwaikhat, H. M. (2024). Optimizing waste management strategies through artificial intelligence and machine learning—An economic and environmental impact study. *Cleaner Waste Systems, 8*, 100158. https://doi.org/10.1016/j.clwas.2024.100158
32. Soni, U., Roy, A., Verma, A., & Jain, V. (2019). Forecasting municipal solid waste generation using artificial intelligence models—A case study in India. *SN Applied Sciences, 1*, 162. https://doi.org/10.1007/s42452-018-0157-x
33. Shekarappa, G. S., Mahapatra, S., Raj, S., & Badi, M. (2023). Smart transportation based on AI and ML technology. In V. Basetti, C. K. Shiva, M. R. Ungarala, & S. S. Rangarajan (Eds.), (pp. 281–299). Elsevier. https://doi.org/10.1016/B978-0-323-99503-0.00018-1
34. Nikitas, A., Michalakopoulou, K., Njoya, E. T., & Karampatzakis, D. (2020). Artificial intelligence, transport and the Smart City: Definitions and dimensions of a new mobility era. *Sustainability, 12*, 2789. https://doi.org/10.3390/su12072789
35. Jemmali, M., Melhim, L. K. B., Alharbi, M. T., Bajahzar, A., & Omri, M. N. (2022). Smart-parking management algorithms in smart city. *Scientific Reports, 12*, 6533. https://doi.org/10.1038/s41598-022-10076-4
36. Thinh, T. N., Le, L. T., Long, N. H., Le Thuc Quyen, H., Thu, N. Q., La Thong, N., & Nghi, H. P. (2021). An edge-AI heterogeneous solution for real-time parking occupancy detection. *Presented at the 2021 International Conference on advanced Technologies for Communications (ATC)*, pp. 51–56. https://doi.org/10.1109/ATC52653.2021.9598291.
37. Szpilko, D., de la Torre Gallegos, A., Jimenez Naharro, F., Rzepka, A., & Remiszewska, A. (2023). Waste management in the smart city: Current practices and future directions. *Resources, 12*, 115. https://doi.org/10.3390/resources12100115
38. ur Rehman, U., Faria, P., Gomes, L., & Vale, Z. (2023). Future of energy management systems in smart cities: A systematic literature review. *Sustainable Cities and Society, 96*, 104720. https://doi.org/10.1016/j.scs.2023.104720

39. PricewaterhouseCoopers. (n.d.). *AI-backed platforms can keep public safer and better informed* [WWW document]. PwC. Retrieved Nov 2, 2024, from https://www.pwc.com/m1/en/media-centre/articles/ai-backed-platforms-can-keep-public-safer-and-better-informed.html
40. Xu, Z., Wang, X., Niu, Y., & Zhang, H. (2022). Robust simulation of cyber-physical systems for environmental monitoring on construction sites. *Applied Sciences, 12*, 10822. https://doi.org/10.3390/app122110822
41. Neo, E. X., Hasikin, K., Lai, K. W., Mokhtar, M. I., Azizan, M. M., Hizaddin, H. F., Razak, S. A., & Yanto. (2023). Artificial intelligence-assisted air quality monitoring for smart city management. *PeerJ Computer Science, 9*, e1306. https://doi.org/10.7717/peerj-cs.1306
42. Mazhar, T., Malik, M. A., Haq, I., Rozeela, I., Ullah, I., Khan, M. A., Adhikari, D., Ben Othman, M. T., & Hamam, H. (2022). The role of ML, AI and 5G technology in smart energy and smart building management. *Electronics, 11*, 3960. https://doi.org/10.3390/electronics11233960
43. Krishnan, S. R., Nallakaruppan, M. K., Chengoden, R., Koppu, S., Iyapparaja, M., Sadhasivam, J., & Sethuraman, S. (2022). Smart water resource management using artificial intelligence—A review. *Sustainability, 14*, 13384. https://doi.org/10.3390/su142013384
44. Roth, Z., & Incera, M. (n.d.). *The rise of AI-powered smart cities* [WWW document]. S&P Global. Retrieved Nov 2, 2024, from https://www.spglobal.com/en/research-insights/special-reports/ai-smart-cities
45. Almaleh, A. (2024). A novel deep learning approach for real-time critical assessment in smart urban infrastructure systems. *Electronics, 13*, 3286. https://doi.org/10.3390/electronics13163286
46. Haghshenas, S. S., Guido, G., Haghshenas, S. S., & Astarita, V. (2023). The role of artificial intelligence in managing emergencies and crises within smart cities. Presented at the 2023 International Conference on Information and Communication Technologies for Disaster Management (ICT-DM), pp. 1–5. https://doi.org/10.1109/ICT-DM58371.2023.10286925.
47. Wu, P., Zhang, Z., Peng, X., & Wang, R. (2024). Deep learning solutions for smart city challenges in urban development. *Scientific Reports, 14*, 5176. https://doi.org/10.1038/s41598-024-55928-3
48. Lungu, M. A. (2024). Smart urban mobility: The role of AI in alleviating traffic congestion. *Proceedings of the International Conference on Business Excellence, 18*, 1441–1452. https://doi.org/10.2478/picbe-2024-0118
49. Rafalia, N., Moumen, I., Raji, F. Z., & Abouchabaka, J. (2024). Elevating smart city mobility using RAE-LSTM fusion for next-gen traffic prediction. *Indonesian Journal of Electrical Engineering and Computer Science, 35*, 503–510. https://doi.org/10.11591/ijeecs.v35.i1.pp503-510
50. Chen, J. (2024). Vehicle detection and tracking in intelligent transportation systems based deep learning. *Applied and Computational Engineering, 33*, 28–33. https://doi.org/10.54254/2755-2721/33/20230227
51. Reddy, S., Pillay, N., & Singh, N. (2024). Comparative evaluation of convolutional neural network object detection algorithms for vehicle detection. *Journal of Imaging, 10*, 162. https://doi.org/10.3390/jimaging10070162
52. Alamsyah, N., Yoga, T. P., & Budiman, B. (2024). Improving traffic density prediction using LSTM with parametric ReLU (PReLU) activation. *JITK (Jurnal Ilmu Pengetahuan dan Teknologi Komputer), 9*, 154–160. https://doi.org/10.33480/jitk.v9i2.5046
53. O'Halloran, J. (n.d.). *ST engineering drives AI smart city transformation in Qatar | computer weekly* [WWW document]. ComputerWeekly.com. Retrieved from Oct 28, 2024, from https://www.computerweekly.com/news/366613637/ST-Engineering-drives-AI-smart-city-transformation-in-Qatar
54. Writer, S. (2024). *Lusail in Qatar taps ST engineering for AI-powered smart city drive*. Frontier Enterprise. Retrieved Oct 28, 2024, from https://www.frontier-enterprise.com/lusail-in-qatar-taps-st-engineering-for-ai-powered-smart-city-drive/

55. Planet Satellite Data, AI Enables Bahrain to Manage Smart Cities. (2024). *Geospatial world*. [WWW Document]. Retrieved from Oct 28, 2024, from https://www.geospatialworld.net/news/planet-satellite-data-ai-enables-bahrain-to-manage-smart-cities/
56. Smart Traffic Solutions to Make Copenhagen Carbon Neutral. (n.d.). *Technolution move*. [WWW Document]. Retrieved Oct 28, 2024, from https://www.technolution.com/move/cases/copenhagen/

Blockchain Framework for Securing Medical Records

K. Geetha, S. Dinesh, S. Aarthi, and N. Sasikaladevi

1 Introduction

Blockchain technology offers a secure, decentralized solution to the problems with conventional medical data sharing and storage techniques, indicating the reason the healthcare sector may benefit greatly from it. The essential concept underlying blockchain technology is its ability to create an immutable, transparent record that ensures the integrity and privacy of highly sensitive information [1–4]. Security and privacy of medical records are important considerations in the centralized architecture of healthcare data management systems. The extensive use of electronic health record (EHR) systems provides a single point of failure and rewards bad actors as information becomes more digitized. On the other hand, a blockchain is powered by a decentralized network of nodes. Every node keeps a precise duplicate of the whole ledger. Due to the decentralized architecture, it is more difficult for unauthorized parties to tamper with sensitive information. As data in a blockchain is irreversible, it is irrevocably recorded and serves as a permanent medical history archive.

Moreover, blockchain technology enhances information transparency, elevating healthcare practitioners' confidence levels. Consequently, patients, healthcare professionals, and other verified individuals can access and validate their medical histories. This transparency in healthcare enables a patient-centered and collaborative approach, while also guaranteeing the accuracy of the information. Smart contracts can significantly simplify the management of medical records. For example, a "Smart Contract"-based management system is straightforward to automate, ensuring that only authenticated individuals can access specific medical records. To ensure the privacy and security of health data using a distributed ledger system in Python, it is likely necessary to establish multiple distributed nodes. For the

K. Geetha · S. Dinesh · S. Aarthi (✉) · N. Sasikaladevi
SASTRA Deemed University, Thanjavur, Tamil Nadu, India

S. M. Idrees et al. (eds.), *Blockchain and Machine Learning Innovations*, Signals and Communication Technology,
https://doi.org/10.1007/978-3-031-88095-7_8

encryption steps needed to check each transaction and add it to the blockchain, each of those nodes would have been used. Furthermore, smart contracts could facilitate the automation of data-sharing agreements and access protocols for the management of medical records. Ultimately, blockchain technology in the healthcare sector has the potential to transform the access, security, and privacy of medical records. Consequently, the risks and vulnerabilities inherent in conventional EMR management systems can be mitigated through the utilization of distributed and immutable ledgers by healthcare application providers. It fosters trust, transparency, and productivity, subsequently resulting in the extensive use of intelligent healthcare systems [15].

1.1 Motivation

The standardization of blockchain-based electronic health record systems is crucial. The absence of standards leads to security and privacy issues and hampers integration with current healthcare systems. Consequently, the establishment and execution of standardization procedures are crucial in addressing these difficulties.

1. ***Scalability:*** The volume of data produced daily may exceed the processing capacity of current blockchain technology. As the demand for EMR administration and storage escalates, it is imperative to develop new, more scalable solutions through research.
2. ***Patient acceptance of blockchain-based electronic health records:*** The patients' limited understanding of blockchain technology and privacy security issues are creating a dilemma. We must resolve privacy concerns to enhance their acceptability and utilization.
3. ***Integration with current healthcare systems:*** The failure to connect blockchain-based electronic health records with existing healthcare systems raises interoperability issues. Establishing protocols is necessary to facilitate seamless system integration and data exchange.

This calls for a process of constant improvement and adapting to new rules and technology. This will make it possible for blockchain-based electronic health records to be used in the long term; models must be made that take all of these things into account.

1.2 Contributions

The suggested "Medichain" guarantees the confidentiality and integrity of healthcare information while expediting the data streamlining process. Through the use of Medichain, a web-based application built in Python, this solution makes the EMR harder to change and makes record maintenance more efficient.

1.3 *Structure of the Article*

The subsequent sections of this work are as follows. Section "Literature Review" examines blockchain technology and its comparison with other pertinent technologies. Section "Results and Discussions" delves into the proposed initiative known as "Medichain." Section "Block Description" provides an overview of the reports and an explanation of the dataset. Section "Results" illustrates the fluctuation in time and memory as the number of blocks in Medichain escalates. Section "Conclusion" provides the conclusion of the paper.

2 Literature Review

Al Mamun and others [1] put forward a novel approach that combines blockchain technology and IPFS to ensure the secure storage of EMR systems. As a distributed file system, IPFS stores Electronic Medical Records (EMR), which maintains a broad network of nodes—certifying that data is robust against modifications and unauthorized interference. On the other hand, the blockchain provides an immutable ledger on which records guarantee permanent auditing of all EMR changes, improving transparency and traceability across the sector. Such a solution maintains the acts of the healthcare sector where EMR confidentiality is enhanced and patients are easily given access.

However, the conventional systems for managing electronic health records raised various concerns; hence, Chelladurai et al. [5] suggested the adoption of new systems based on blockchain to automate electronic health records. It allowed smart contracts in the blockchain domain to securely store, manage, and exchange electronic health records (EHRs). This project will provide great advantages like unchangeable records of the patients, secure and easy updates of medical records, easy sharing of health information, patient-centric viewing rights, and improved security and integrity. The approach the authors propose is aimed at changing the operations of EHR systems by enhancing patients' rights, better managing healthcare services, and preventing data security and integrity.

Privacy safeguarding measures have been integrated into designing an electronic health record management system based on a blockchain platform by Karunakaran et al. [6]. From this proposal, it is expected that patients' information will be kept secure with the help of a technological solution involving the blockchain model, and patients will remain confidential since they control the health record. This allows patients to keep their EMRs, indicating that healthcare professionals often give consent or deny access to their information. Data integrity protection and management of patient information is enhanced by the use of searchable encryption and blockchain technology as offered in the framework by Tang et al. [7]. This approach allows for efficient searching of keywords in encrypted medical records, using searchable encryption, while ensuring the protection of individual privacy.

Blockchain technology is utilized to protect the system for storing encrypted search indexes and verification indexes thus protecting them against modification and ensuring their tracking. The program that has been proposed will provide individuals with tools that would enable them to have control over all their medical data, along with allowing sharing but with limitations as well as ensuring both security and transparency of the records of the medical practice history. Azbeg et al. [8] analyze a chronic illness self-management system based on blockchains that provide security for the exchange of medical data for chronic diseases that require long-term monitoring.

The scheme combines proxy re-encryption-based smart contracts and achieves data authenticity, patient privacy, and access control. It makes use of a large medical file that includes an InterPlanetary File System and blockchain to ensure that there is an unchangeable record of the time of access and any alterations made. This system has the potential to overcome challenges that come with the security and privacy of existing medical data management systems which may lead to better patient care and management of diseases. According to Lin et al. [9], an efficient model for the context of the exchange of electronic medical records on the blocks was developed with built-in protection against malevolent dispersion and the ability to suppress altered electronic medical records. It incorporates a new Universal Designated Verifier Signature Proof technique based on the Elliptic Curve Digital Signature Algorithm for more efficiency, but less computational resources.

Smart contracts would be used to securely store, manage, and disseminate EHR. The authors pointed out the benefits of EHR integration with blockchain technology as increased security, improved patient self-management, and better collaboration amongst actors within the healthcare structure. It can help in building a secure, transparent, patient-centric, and more effective and efficient healthcare system, which counters some of the challenges of the existing EHR systems. MedChain is a proposed framework by Darghmi et al. [10] for managing medical records and access permissions using blockchain technology.

MedChain offers patient-centered access control on existing gaps in current medical record systems where patients are alienated in their care by making access safe and collaborative. It employs smart contracts, establishing strict access controls and an immutable audit log of every access and modification made. We also secure patient data using encryption and potentially implement an incentive structure for healthcare providers who maintain accurate and up-to-date patient records. Under this arrangement, healthcare records stand to be more effective, open, and highly secure. The proposed system can make the management of medical records more effective, transparent, and secure. A medical chain is a solution that implements EMR systems as smart electronic health records with blockchain technology to correct the deficiencies of conventional EMRs as noted in the study [11].

Medical Chain utilizes smart contracts that allow data to be encrypted while a complete audit trail of access and change is retained in an unmodifiable format. Moreover, it applies access limitations in a controlled manner. It even incorporates incentive mechanisms for healthcare providers to make them maintain the accuracy and currency of information within the system. This would promote efficiency, security, and transparency in the management of medical records. Raghav et al. [12] also proposed a secure and privacy-preserving blockchain-based system for healthcare data management.

The framework ensures both the data's integrity and the patient's privacy through the fusion of blockchain technologies coupled with other cryptographic techniques. The patient's activity in accessing EMR is illustrated in Fig. 1. The framework introduces a unique technique for data sanitization and restoration, ensuring minimal privacy infiltration and effective medical data analysis. Moreover, this framework allows legitimate users to exchange health data safely and responsibly to improve care coordination for patients. The architecture that has been proposed is capable of revolutionizing the management of healthcare data by overcoming the security and privacy concerns that come with the traditional approach. Tan et al. [13] offered a blockchain-based solution for the secure and confidential exchange of the COVID-19 medical database.

The use of ciphertext-policy attribute-based encryption ensures that only authorized users can view the health record. Some extra system parameters include the public keys and the revocation lists, which are stored on the blockchain so that a coherent audit trail is preserved and all the information is visible. This proposed methodology can alleviate the security and privacy issues surrounding the transmission of medical information about COVID-19 while at the same time improving

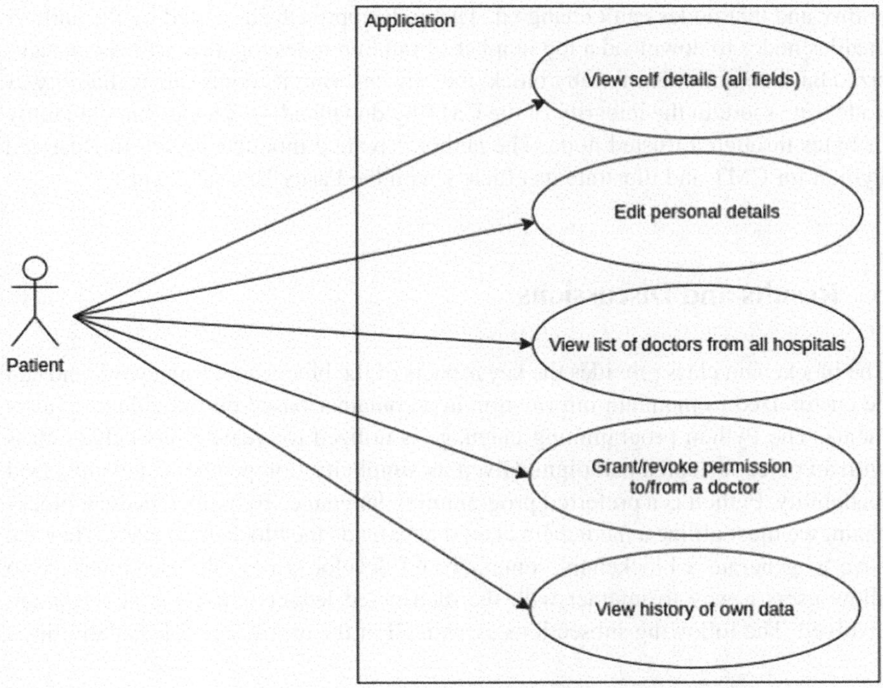

Fig. 1 Activity diagram of patient accessing EMR

research and the care of patients during the pandemic. Shahnaz et al. focused on the problems of conventional EMR systems in their research [14]. Apart from listing the problems, they also highlighted the benefits of integrating blockchain into electronic health records management, such as increased security, patient-centeredness, and better coordination. Several researchers have examined blockchain-based EMRs' storage, access control, auditing, and pros and cons.

The participants in proof-of-work systems are the depth miners, responsible for finding the solutions to the particular equations. To add a new block to the blockchain, the miner needs to append the block's hash to a unique number called a nonce, such that the value of the appended number remains below some threshold. This is a very resource-consuming process and requires a lot of computational power. If the depth puzzle is varied and shifted, it will be possible to ensure that blocks are deployed at a time interval and with a target number. After he receives the solution to the problem, the miner broadcasts it and gets it confirmed by other miners. The temptation for bad players is removed because to change the block, it must be proven that the required amount of work was spent on the proof of work of that block and all blocks after it, which is impossible in a normal scenario. In blockchain networks, PoW facilitates reliable and unrelated consensus. It is shown that a coded Merkle tree, which is used to build a new kind of hash accumulator for blockchains, can protect against data availability attacks at a constant cost.

In the creation of CMT, sparse erasure codes are used, and peeling decoding is used as a recovery method to make sure that the data is quickly checked for availability and that nodes can't change it. The node approach suggested by the authors enables nodes to download a log number of random bytes together with a constant-sized hash commitment for any block, thereby ensuring its complete availability. A node can ascertain the integrity of the CMT by downloading a logarithmic quantity of bytes through a trusted node. The authors create a modular library in Rust and Python for CMT and illustrate its efficacy with the Parity Bitcoin client.

3 Results and Discussions

The blockchain class provides the key aspects of the blockchain framework and can be customized to maintain information in a container based on individual requirements. The Python programming language is utilized to create a blockchain class with an object-oriented paradigm. Given its simplicity, compactness, and enhanced readability, Python is a preferred programming language. To build a medical blockchain, we must utilize a medichain class that extends the blockchain class. Then we have to generate a blockchain container and develop a console user interface to allow users a way to interact with the distributed ledger where the subtasks are involved. The following subsections explain all of the user-accessible capabilities.

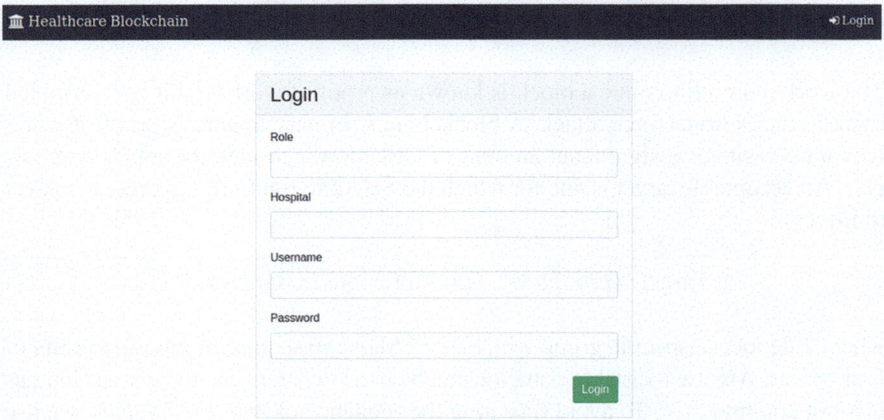

Fig. 2 GUI designed for Medichain Framework

We subject each record to rigorous security protocols to protect it against unauthorized access and tampering. An essential element of this security architecture is that each record will own a distinct hash value. The hash value serves as a distinctive digital fingerprint of the record data and cannot be recalibrated. It featured a highly efficient hashing technique that ensured a balance between cryptographic strength and computational performance. Given that hash value generation is relatively inexpensive, it facilitates rapid and straightforward record addition via the GUI.

We created a web application as in Fig. 2 that enables the concurrent uploading of patient hospital records into the EMR system. The proposed method ensures the confidentiality and authenticity of inserted data while enhancing speed relative to existing studies. Additionally, we have implemented a blockchain-inspired architecture in Python to fortify the EMR against various assaults targeting data manipulation, and we have performed several experiments to validate the efficacy of our design. We have also made the hospital administration system more trustworthy by using blockchain ideas, such as cryptographic hashing. This shows that the system is dedicated to keeping data safe and real.

3.1 Building a Container for Blockchain

A new class named Medichain (Blockchain) was established to store medical records in a Blockchain block. The Blockchain class was created independently and contains member methods that provide the core functionalities of a Blockchain. Section "Block Description" reviews these member functions and other important characteristics in-depth.

3.2 Proof of Work

The work required to mine a block is known as proof of work, and it is determined by initiating a brute-force attack. A blockchain's primary feature is proof of work, which necessitates a significant amount of GPU power to identify appropriate targets. An acceptable target is one for which the SHA256 run finds the precise answer to Eq. (1).

$$\text{Target} = \text{proofnew}2 - \text{proofprevious}2 \vee \text{target} \in T \tag{1}$$

where T denotes a specific group with hashed SHA256 sequences having a prefix of four zeroes. Any two-variable equation can be used to determine the goal as long as it is not commutative. To avoid repeating the sequence of two proof values, we use non-commutative equations. We could have gotten P as the next proof for the third part of the series if we had chosen an equation like $P + Q$, with P being the current proof and Q being the next proof. This is because $P + Q$ would have satisfied the target equation, which is $P + Q = Q + P$. This pattern can be continued indefinitely.

3.3 Verifying the Legitimacy of Blockchain

If any of a blockchain's blocks fail to meet specific criteria, it becomes untrustworthy and suggests tampering with its contents. Every blockchain block must meet the two requirements listed below, which must be met by every blockchain block for a legitimate connection to be present inside the sequence of blocks.

$$\text{ofpreviousblock} = \text{SHA256}(\text{previousblock})$$

$$\text{Target} = \text{proofnew}2 - \text{proofprevious}2 \vee \text{target} \in T$$

A subset of SHA-256 hashes using 0000 as a prefix is called SHA256(target) T.

Computational Complexity: Order O(P) is the time complexity, where P denotes the count of blocks in the Medichain. Order O (1) is the auxiliary space complexity. The method in sections "Building a Container for Blockchain, " "Proof of Work," and "Verifying the Legitimacy of Blockchain" evaluates each value to see if the target characteristics and prior hash of a block in a blockchain meet the requirements mentioned above.

3.4 *Accessing Blockchain through a Command Line Interface (CLI)*

CLI provides the following options for simple blockchain interaction, and it may perform numerous operations.

1. Create a new Blockchain with Blockchain(): The genesis block blockofprevious-block = '0' is generated. The option to generate a genesis block, the leading edge of a blockchain, will initiate it.
2. Construct a block in a Blockchain: To add a new block to the Blockchain (evidence, prior hash), select the create block option.
3. View the contents by going to a blockchain: Examine each block individually on the blockchain using operations such as print Block_index, print Block_time-stamp, print Block_proof, and print Block_records. By selecting these options, the Blockchain's structure, which includes the Block index, Block timestamp, and print, will manifest as an unbreakable JSON object.
4. Check and certify the blockchain: Choosing this option will start the validation process for each block, as outlined in section "Verifying the Legitimacy of Blockchain".

3.5 *Merkle Tree*

3.5.1 The Merkle Tree Creation

To create the Merkle tree, first, we have to break the data into manageable pieces or sections. A leaf node is created by hashing each data block or chunk separately to get a unique hash value. The hash values represent the leaves of the Merkle tree [16]. Subsequently, pairwise hashing must be employed to hash the combined values of two adjacent leaf nodes. The generated hash values are the parent nodes of the subsequent level.

Repetition of Pairwise Hashing: We must keep hashing pairs up the tree until there is only one remaining root hash. This root hash represents all of the data in it.

3.5.2 Checking the Merkle Tree

To validate the Merkle tree, we need to determine the hash value of a specific data block or chunk, which will confirm the identification of the leaf node hash. The following choices will be made to verify the Merkle tree.

- *Locate Matching Leaf Node:* In the Merkle tree, we have to determine the location of the leaf node that corresponds to the hash value.
- *Trace Ancestral Nodes:* Compile the hashes as we ascend the parent nodes' journey, beginning at the leaf node.
- *Compare Accumulated Hash:* We have to verify that the accumulated hash value matches the Merkle tree's root hash. The data block or chunk is regarded as validated if they match.

4 Block Description

4.1 Dataset Description

The collection of data utilized to simulate the built program includes data entries of 141 patients, each having a unique disease. The following fields are present in each dataset entry. (Patient's ID, Patient's name, Surname, Diagnose Code, email, Gender, Telephone, Date of birth, International Classification of Disease-ninth revision, Nation of origin, Residence, Enrolled Date, SicknessId).

4.2 Block Description

The following fields can be observed in a Blockchain block.

1. Index: the block's index number in the Blockchain.
2. Merkle root: SHA256 hash root of Merkle tree.
3. Previous hash: SHA256 (# of previous block).
4. Consensus: proof of work (Consensus protocol).
5. Records: a collection of patient records (the maximum allowable records per block is 10; however, this can be readily changed if necessary).

5 Results

Figure 3 illustrates how the two fundamental characteristics of the Blockchain change as the number of blocks increases. The previously described dataset has been simulated in the Figure 4. The graph shows that the average execution time increases linearly as the number of blocks increases. The following plot aggregates the duration of check chain validity and mine block operations, subsequently plotting the resulting graph against the number of blocks. Execution time of both these operations increases linearly, and it is clear that for all the cases, check chain

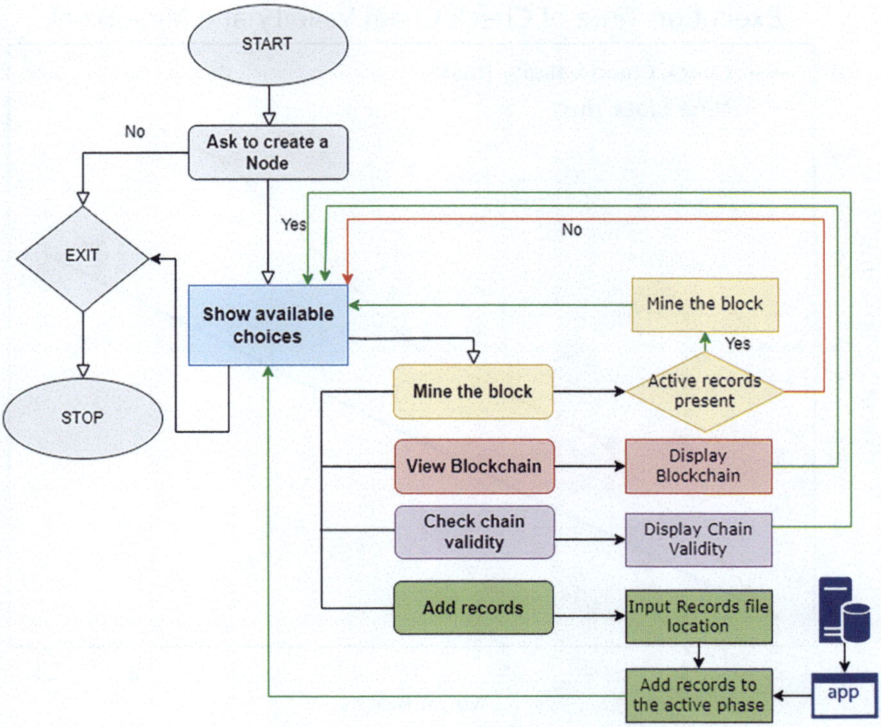

Fig. 3 Flowchart of the Medichain command line interface

validity has always been obtained after consuming more time than required for mining a block. This refers to a situation where when the blockchain is increasing in size, validation of the chain is computationally intensive; hence, it may affect the system's performance and scalability (Fig. 4).

The graph in Fig. 2 illustrates shows how the system in question works and how the number of blocks and the amount of memory used (in bytes) affect throughput and latency. We realize that as the number of blocks increases, the memory size also increases, potentially leading to system degradation. First, between 2 and 4 blocks, the increase in memory usage is observed, indicating that it has a negligible effect on the throughput and latency. Nevertheless, the steep rise of memory usage at blocks 4–6 means that the throughput is likely to decrease with latency growing up, probably due to storing data or metadata with larger sizes in those blocks. After block 6, there is an identifiable growth of memory usage, which indicates much more stability as a system in terms of predictable performance as indicated in the (Fig. 5). This trend proves the need to work on memory management so that such systems can achieve high throughput and low latency and, as a result, be scalable in the application of blockchain to large datasets, such as the healthcare sector. Furthermore, there is a linear relationship between memory utilization and block

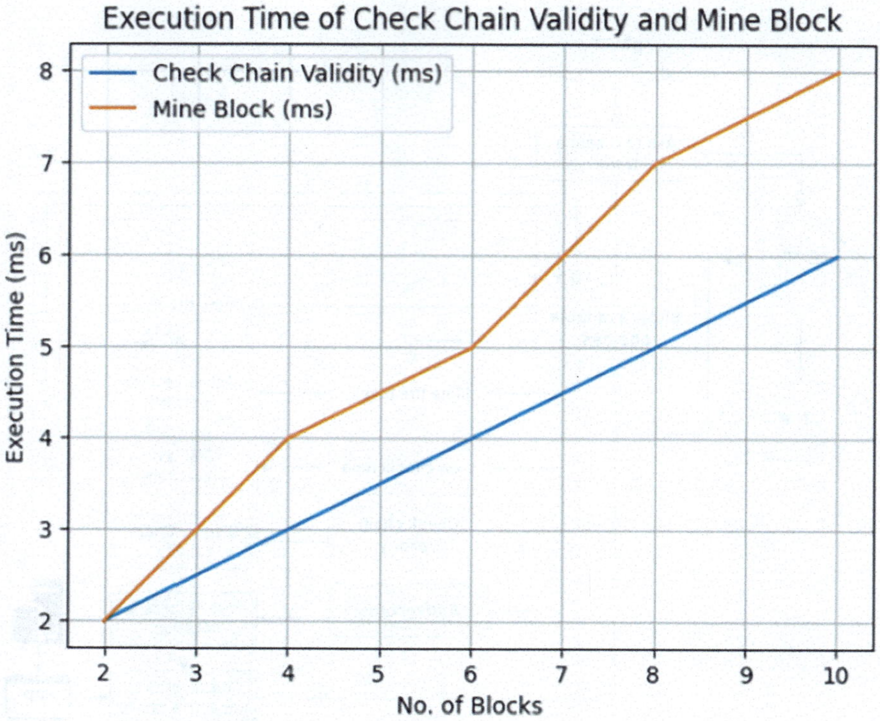

Fig. 4 Variation in time as the number of blocks increases

count. Incorporating the blockchain-based Medichain framework has effectively solved the major problems with managing healthcare data. Key findings from the study are as follows:

1. **Enhanced Data Security and Integrity:** By using blockchain's immutable ledger, the authenticity of the healthcare records has been protected so that outsiders can't change them and stakeholders can trust the records. During the testing phase, we reported no data breaches or unauthorized access incidents, demonstrating the effectiveness of the framework.
2. **Improved Interoperability:** Medichain showed interoperability with various EMRs; hence, sharing data across different providers was smooth and secure. Smart contracts' employment enabled the automated release and sharing of records, which otherwise would require the input of many people and much time.
3. **Patient Empowerment:** Patients were given the ability to customize and control their health data through permissioned access, meaning that patients were fully in control of their records. Feedback indicated an increased level of confidence and satisfaction with the system.
4. **Performance Metrics:**

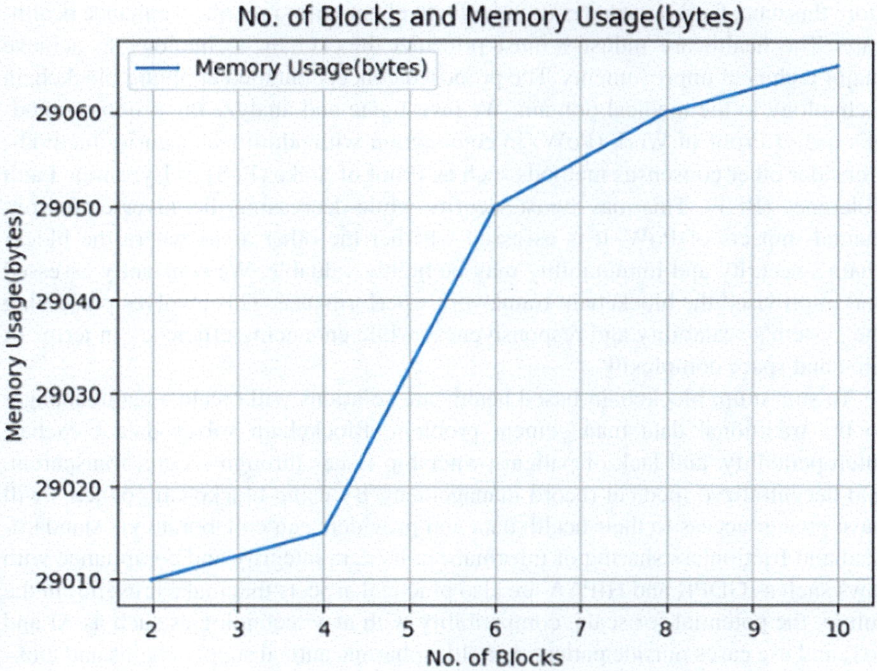

Fig. 5 Variation in memory use as the number of blocks increases

(a) *Latency:* Average data retrieval times did not increase even with higher transactions to afford quick and near real-time access to important records.
(b) *Scalability:* This combination of three layers of the blockchain rendered the solution efficient and able to handle large datasets needed for future large-scale healthcare practices.
(c) *Compliance and Auditability:* The framework addresses a few regulatory measures, including GDPR and HIPAA, and maintains records of all the interactions that occur with data.
(d) *Cost Efficiency:* The decentralized system eliminated overhead costs for processes such as data validation and access control, improving the overall cost for healthcare providers.

6 Conclusion

Many industries stand to gain drastically from the application of blockchain technology. It could make the existing systems highly secure and impenetrable. Furthermore, data is expanding rapidly in certain industries, such as healthcare. Consequently, there is a high demand for technologies such as blockchain to safely store this data, facilitate analysis, and effectively monitor records to enhance healthcare. The healthcare industry must prioritize blockchain technology to achieve major technical improvements. The proposed concept entailed applying blockchain technology to the medical domain. We investigate and analyze the proposed model's use of Proof of Work (PoW) in conjunction with additional security methods. Consider other consensus methods such as Proof of Stake (PoS) or Byzantine Fault Tolerance (BFT). This may boost security while decreasing the adverse environmental impacts of PoW. It is assessed whether the other areas where the blockchain's security and immutability may be highly valuable. We constantly assessed and improvised the blockchain framework's performance. This involves preserving the system's scalability and responsiveness while enhancing efficiency in terms of time and space complexity.

To sum it up, blockchain-based healthcare solutions will create a game-changer to the traditional data management problem. Blockchain solves data breaches, interoperability, and lack of patient-ownership issues through secure, transparent, and decentralized medical record management. By using blockchain, patients will have greater access to their health data and providers can collaborate via standardized and frictionless sharing of information. Its data integrity and compliance with laws such as GDPR and HIPAA are also practical aspects that make it useful. In the future, the potential for scale, compatibility with new technologies such as AI and IoT, and use cases outside patient data (like pharmaceutical supply chains and clinical trials) make blockchain an innovation platform in the healthcare space.

Acknowledgments This part of this research work is supported by the Department of Science and Technology (DST), Science and Engineering Board (SERB), Government of India under the TARE grant (TAR/2020/000259).

References

1. Al Mamun, A., Faruk Jahangir, M. U., Azam, S., Kaiser, M. S., & Karim, A. (2020, December). A combined framework of interplanetary file system and blockchain to securely manage electronic medical records. In *Proceedings of International Conference on Trends in Computational and Cognitive Engineering: Proceedings of TCCE 2020* (pp. 501–511). Springer Singapore.
2. Idrees, S. M., Nowostawski, M., Jameel, R., & Mourya, A. K. (2021). Privacy-preserving infrastructure for health information systems. In *Data protection and privacy in healthcare* (pp. 109–129). CRC Press.
3. Rather, I. H., & Idrees, S. M. (2021). Blockchain technology and its applications in the healthcare sector. In *Blockchain for healthcare systems* (pp. 17–25). CRC Press.
4. Idrees, S. M., & Nowostawski, M. (Eds.). (2024). *Blockchain Transformations: Navigating the Decentralized Protocols Era*. Springer.
5. Chelladurai, U., & Pandian, S. (2022). A novel blockchain based electronic health record automation system for healthcare. *Journal of Ambient Intelligence and Humanized Computing, 13*(1), 693–703.
6. Karunakaran, C., Ganesh, K., Ansar, S., & Subramani, R. (2021). A privacy preserving framework for health records using blockchain. *Pertanika Journal of Science & Technology, 29*(4).

7. Tang, X., Guo, C., Choo, K. K. R., Liu, Y., & Li, L. (2021). A secure and trustworthy medical record sharing scheme based on searchable encryption and blockchain. *Computer Networks, 200*, 108540.
8. Azbeg, K., Ouchetto, O., & Andaloussi, S. J. (2022). Access control and privacy-preserving blockchain-based system for diseases management. *IEEE Transactions on Computational Social Systems., 10*(4), 1515–1527.
9. Lin, C., Huang, X., & He, D. (2023). Efficient blockchain-based electronic medical record sharing with anti-malicious propagation. *IEEE Transactions on Services Computing., 16*(5), 3294–3304.
10. Daraghmi, E. Y., Daraghmi, Y. A., & Yuan, S. M. (2019). MedChain: A design of blockchain-based system for medical records access and permissions management. *IEEE Access, 7*, 164595–164613.
11. Retrieved from https://medicalchain.com/Medicalchain-Whitepaper-EN.pdf
12. Raghav, N., & Bhola, A. K. (2022, January). Secured framework for privacy preserving healthcare based on blockchain. In *In 2022 International Conference on Computer Communication and Informatics (ICCCI)* (pp. 1–5). IEEE.
13. Tan, L., Yu, K., Shi, N., Yang, C., Wei, W., & Lu, H. (2021). Towards secure and privacy-preserving data sharing for COVID-19 medical records: A blockchain-empowered approach. *IEEE Transactions on Network Science and Engineering, 9*(1), 271–281.
14. Shahnaz, A., Qamar, U., & Khalid, A. (2019). Using blockchain for electronic health records. *IEEE Access, 7*, 147782–147795.
15. Tanwar, S., Parekh, K., & Evans, R. (2020). Blockchain-based electronic healthcare record system for healthcare 4.0 applications. *Journal of Information Security and Applications, 50*, 102407.
16. Retrieved from https://cointelegraph.com/explained/proof-of-work-explained

Mapping Machine Learning Trends in Predictive Analytics for Alzheimer's Disease

Roshan Jameel, Alisha Sikri, Sheikh Mohammad Idrees, Bhawna Wadhwa, Vikas Sagar, and Benita Christopher

1 Introduction

Machine Learning (ML) is a branch that comes under Artificial Intelligence, which is basically the training of the programs through the dataset such that they can learn the inferences and predict the new data based on their learning [1]. These algorithms have the capability to learn and adjust by themselves to improve their performances in an automatic manner [2, 3]. There are several machine learning techniques such as support vector machine (SVM) that works good with small dataset and is capable of classifying in a supervised manner [4, 5]. Logistic regression is another approach used for prediction of association of dependent variable with the independent variable [6]. Random forest comes under supervised machine learning techniques that is very adaptable and simple yet gives outstanding results [7]. One more machine learning technique used for the classification problems is the decision tree that gives consistently accurate results [8]. Several other machine learning techniques such as K-nearest neighbor (KNN), linear discriminant analysis (LDA), and Naïve Bayes are also being used for the exploration of data to extract useful information and predictive analysis. Machine learning has proved its effectiveness in multiple domains including healthcare. It is used for prediction of diseases [9–11, 88–90],

R. Jameel (✉) · B. Christopher
Westford University College, Sharjah, UAE

A. Sikri
JIMS, Rohini, Delhi, India

S. M. Idrees
Department of Computer Science, Norwegian University of Science and Technology, Gjøvik, Norway

B. Wadhwa · V. Sagar
Noida Institute of Engineering and Technology, Noida, India

S. M. Idrees et al. (eds.), *Blockchain and Machine Learning Innovations*, Signals and Communication Technology,
https://doi.org/10.1007/978-3-031-88095-7_9

pharmaceutical supply chain [12, 13], health insurance frauds [14], etc. This paper analyzes the articles published up to the year 2022 that use machine learning techniques for prediction of Alzheimer's disease.

Alzheimer's disease is an irreversible illness that affects the brain of the patient by breaking the tissues with time and making it impossible for the person to perform normal day-to-day activities [15, 16]. It usually occurs to old people who are of age more than 65 years [17]. It affects millions of people worldwide. The symptoms of this disease are improper sleeping habits, anxiety, difficulty in performing everyday tasks like eating, reasoning, writing, identifying other people, etc. These symptoms are usually mild in the beginning and become prominent, after around 10 years of the commencement of the damage of the brain, which progresses to complete memory loss [18–20]. It comes under the category "Dementia," which is generally reported in elderly people. So far, Alzheimer's disease has no cure [21]. However, some studies suggest that its symptoms can be delayed by therapies if detected at an earlier stage [22]. This is where machine learning algorithms might help as they can predict diseases with good accuracy. The electronic health records (EHRs) of the patients at risk of developing dementia can be collected in the form of laboratory results, vitals, medicines, genetic information, etc. [23, 91], and the machine learning analysis can be performed on the collected data. In some of the cases, the doctors recommend specific tests for such as "magnetic resonance imaging" (MRI), "position emission tomography" (PET), "single photon emission computed tomography" (SPECT), and cerebrospinal fluid (CSF) [24].

The research papers published for prediction of Alzheimer's disease using machine learning has speeded up in the past 5 years, and it has become a topic of interest for researchers from various fields. This article presents a detailed analysis of the papers, reviews, book chapters written in this field by various authors from countries across the world. The chapter is organized in the following sections: Section 2 discusses about the machine learning techniques and modalities used in the field of Alzheimer's Disease prediction followed by Sect. 3 that mentions about the methodology opted for conducting the analysis, Sect. 4 depicts the various findings and their interpretations achieved during the analysis, finally Sect. 5 summarizes the findings and concludes the analysis done in the paper.

2 ML for Alzheimer's Disease

The machine learning techniques are of two types: Supervised and Unsupervised. To implement supervised machine learning techniques, a labeled dataset is required in which the input and corresponding output values are provided so that the training of the algorithms can be done, and the predictions can be made with accuracy and efficiency. In healthcare domain, the availability of such types of databases is not always possible [58]. However, there are some databases that are publicly available for studying and analyzing the Alzheimer's disease are "Alzheimer's Disease Neuroimaging Initiative" (ADNI) [25], "Australian Imaging Biomarkers & Lifestyle

Table 1 Publicly available Datasets

Dataset name	Description of dataset
ADNI-1	200 subjects with "Normal Controls," 400 with "Mild Cognition Impairment," and 200 with early "Alzheimer's Disease"
ADNI-GO	ADNI-1 database along with 200 subjects with early "Mild Cognitive Impairment"
ADNI-2	ADNI-1 and ADNI-Go along with 150 subjects with "Normal Controls," 100 subjects with "early Mild Cognitive Impairment," 150 with "late Mild Cognitive Impairment" and 150 with "early Alzheimer's Disease"
ADNI-3	ADNI-1, ADNI-Go, and ADNI-2 along with 133 subjects with "Normal Controls," 151 subjects with "Mild Cognitive Impairment," and 87 with "Alzheimer's Disease"
OASIS-1	316 subjects without "dementia" and 100 with "dementia"
OASIS-2	72 non-demented subjects and 64 demented
OASIS-3	609 subjects with "Normal Controls" and 489 with different stage of "cognitive impairment"
AIBL	211 subjects with "Alzheimer's Disease," 133 with "Mild Cognitive Impairment," and 768 with "Normal Controls"

Flagship Study of Imaging" (AIBL) [26], and "Open Access Series of Imaging Studies" (OASIS) [27]. Whereas some of the widely used software for processing the data are Statistical Parametric Mapping (SPM) [28], Freesurfer [29], etc. Table 1 below gives a brief about the most prominently used databases.

2.1 Support Vector Machine

The Support Vector Machine (SVM) is a machine learning-based algorithm that is used for problems associated with classification and regression [59]. The aim of this technique is to locate a hyperplane that can differentiate and classify between the various classes of the data points with maximum margins. In case of non-linear data points, the SVM extends by using the kernel functions. There is numerous research that used SVM for predicting Alzheimer's Disease. Most of the work done in this field has used the dataset consisting of MRI images; however there are other data types of PET, SPECT, etc. that are also being used by the researchers. SVM and decision tree algorithm was combinedly used in [60] on the dataset with MRI for classifying the subjects having Alzheimer's Disease, Mild Cognition Impairment, and Normal conditions. A computer-based system was designed by [61] for diagnosing Alzheimer's disease by using the Independent Component Analysis techniques for extracting the features and SVM for classifying the subjects. Whereas the SVM was implemented in [62] for classifying Alzheimer's Disease and Mild Cognition Impairment from MRI. Further, they have implemented a Particle Swarm Optimization (PSO) technique for optimizing the SVM parameters. In [63], the fMRI data was used to classify the subjects into Alzheimer's Disease and Normal

Controls. The authors proposed an SVM-based technique that selects the data samples and features in a random manner and create multiple SVMs. The authors of [64] used the SVM technique to calculate the prediction accuracy of Alzheimer's disease for different types of data like MRI, PET, and SPECT. They concluded that none of the modalities was found to be superior over other. In [65], the authors worked on MRI and PET images to classify the subjects into Normal Control, Alzheimer's Disease, converted Mild Cognitive Impairment (cMCI), and stable Mild Cognitive Impairment (sMCI).

2.2 *Random Forest*

Random Forest is a tree-based ensemble machine learning technique works on the concept of decision trees in which training of multiple trees is done parallelly. The features are selected after every iteration for individual trees and the results are aggregated. The training is done by taking different set of features every time. RF has proved its efficiency in prediction of Alzheimer's disease, and some of the research works are discussed in the following paragraph. The MRI images are used in [66] for diagnosing the presence of Alzheimer's disease. The authors generated the histograms of the images and applied several machine learning algorithms including RF to identify the subjects that have Alzheimer's Disease. They concluded that the RF outperformed all the other algorithms. The researchers [67] applied 25-fold RF on MRI data from ADNI and AIBL databases. They compared the performance of RF on single datasets and combined datasets and concluded that better results were achieved when the combined dataset was used. The PET images were used by the authors in [68] and the features were extracted and identified the patients with Mild Cognition Impairment using both RF and SVM. A review of various RF-based algorithms used to classify Alzheimer's disease is presented in [47]. As mentioned above, most of the analyses done in this field are performed on image data; however, the demographic and genetic data was used, and the diagnosis is using RF [69].

2.3 *Deep Learning*

Deep learning is a technique that has been used in the healthcare industry for detecting Alzheimer's disease [76]. There are several deep learning-based techniques such as ensemble deep learning [70, 71], convolutional neural network (CNN) [72, 73], recurrent neural network (RNN) [74], and deep belief network [75]. CNN consists of an architecture that is deep and consists of multiple connected layers. The depth of these layers depends on the type of architecture used. The CNN models are trained on image datasets. The widely used datasets for implementing CNN in Alzheimer's prediction are Cifar, ImageNet, etc. There are several review articles that discuss the different deep learning techniques for diagnosing Alzheimer's Disease [76, 77]. CNN and RNN were implemented on image data for diagnosing two neurological diseases Alzheimer's disease and Mild Cognition Impairment [77]. The authors of [78] used datasets of different modality like MRI data, demographic data, cognitive data, and CSF for predicting the development of Alzheimer's disease using RNN. A six-layer architecture was implemented in [79] that consisted of three CNN layers and three fully connected layers.

2.4 K-Means Clustering

It is an unsupervised machine learning algorithm that divides the data into k number of clusters that has nearby mean value. This is a traditional clustering technique that is simple in implementation and has been used in multiple applications. It has also been implemented in the diagnosis of Alzheimer's Disease prominently for segmenting the various regions of the brain for a pre-classification step. However, there is a limitation of this algorithm that the value of "k" is supposed to be known beforehand. In [80], the researchers performed the clustering the MRI data into different clusters based on the various regions of the brain. In this work, the k-means algorithm is used for clustering the parameters and Swarm Optimization technique to segment. The authors of [81] applied machine learning on MRI dataset for classifying the Alzheimer's Disease, Mild Cognition Impairment, and Normal Conditions group. The features were extracted after segmenting the brain into 3*3*3 patches. In [82], the k-mean and k-medoids algorithms were implemented to cluster the features and identify the Alzheimer's Disease and Normal Conditions group. As per their research, the k-medoid was giving better results as compared to k-means algorithm. K-means clustering is applied for identification of sub types of Alzheimer's Disease using Electronic Health Records (EHRs) [83].

A summary of some other relevant references along with the database used, machine learning technique implemented, and the data type is shown in Table 2 below.

Table 2 Summary of Relevant Literature

Reference	Database used	Machine learning technique implemented	Imaging technique used for data collection
[84]	ADNI	Ensemble Algorithms (Deep Belief Network)	PET/MRI
[85]	ADNI	Hidden Markov Model	MRI
[86]	ADNI	2D Convolutional Neural Network	MRI
[67]	ADNI and AIBL	Random Forest	MRI
[87]	ADNI	2D Convolutional Neural Network	MRI
[88]	ADNI	Support Vector Machine	Protein sequence information
[66]	ADNI	Random Forest	MRI
[89]	MIRIAD	Convolutional Neural Network, Support Vector Machine, Random Forest, K-Nearest Neighbor	MRI
[79]	N/A	Convolutional Neural Network	MRI
[90]	N/A	Support Vector Machine, Adaptive Neuro Fuzzy Inference System, Artificial Neural Network	rsfMRI, DTI

3 Data Collection and Selection

In this section the data collection and selection criteria is discussed along with the analysis performed on the chosen data.

3.1 Data Source

To conduct this analysis, the data was collected from Scopus and a total of 3182 papers were retrieved. Scopus was chosen as a source to extract information as it consists of abstracts and citation details of the articles published in several journals, conferences, and book series from more than 11,000 publishers [30]. It covers data from around 240 disciplines [31]. The other databases available are ACM libraries, IEEE, Web of Science, etc. However, the database offered by Scopus contains the articles published in ACM and IEEE, and the API it offers is simplified and flexible as compared to the Web of science [32].

3.2 Criteria Used for Retrieving Articles

Since we wanted to conduct the analysis of Alzheimer's disease using machine learning techniques, initially the papers were selected using "Alzheimer's Disease" and "Machine Learning" as searching criteria that were written in "English" language. The dataset was analyzed with all the 3182 retrieved publications together, and then the search was refined to extract "research articles," "review articles," and "book chapters" individually.

3.3 *Meta-Data Extracted*

The following data was extracted from the Scopus for each of the articles:

Citation Information: (1) Author's Names, (2) Author's ID, (3) Title of Document, (4) Year of publication, (5) Source from where it has been extracted, (6) Number of citations
Bibliographical Information: (1) Affiliations of the Authors, (2) Publisher's name
Abstract and Keywords: (1) Abstract, (2) Author keywords, (3) Index keywords
Miscellaneous: (1) References

3.4 *Analysis Conducted*

After retrieving the papers, the analysis was done on the basis of various criteria such as the number of papers published per year, the areas in which these researches are done, the global fora that publishes the articles in this domain of research, the top authors in the field with maximum number of publications and maximum number of citations, most frequently used keywords, top countries contributing in the research. The top 20 most cited articles, top 20 most cited review papers, and top 15 most cited book chapters are also retrieved. Furthermore, a bibliometric analysis is also done to extract various relationships and links among different aspects of the research.

3.5 *Co-occurrence Analysis*

Analysis is done to understand and investigate the underlying relationship among various bibliometric elements such as co-occurrence of keywords, co-occurrence of authors (authors, country, organization), and co-occurrence of the cited references. The co-occurrences are depicted in the form of graphs which are created using a software called VOSviewer.

4 Analysis and Interpretation

The analysis of the retrieved data and the results observed are discussed and interpreted in this section.

4.1 Publications and Fora

The first ever paper in the field was written in the year 1996, and the number of publications is increasing every year as more researchers are working in this field. As depicted in Fig. 1 below, the count of publications is continuously growing every year and it can be seen, the publications count has seen a sudden growth in the past 5 years. There were only 217 articles published in 2018 which has increased to 907 articles 2022, which is more than 400% of growth.

There are more than 150 Scopus indexed publishers worldwide that has published research articles in the field. The top ten publishers with maximum number of publications are depicted in Fig. 2 below.

4.2 Most Common Keywords

There were around 160 keywords used by the authors in the extracted articles. A little pre-processing was done on the keywords manually. The name of the disease is used as a keyword with two different spellings, i.e., "Alzheimer Disease" and "Alzheimer's Disease." During the analysis, the results of both the keywords are combined and found to be the most frequently used keyword. Same is done with the second most frequent keyword "Human/Humans" as it was used in singular and in plural in different publications. The third most frequently used keyword was found to be "Machine Learning" followed by other keywords like neurodegenerative disease, aged, etc. as depicted in Fig. 3 below. The other keywords are not shown in the figure, as their count was very less in comparison with the depicted ones.

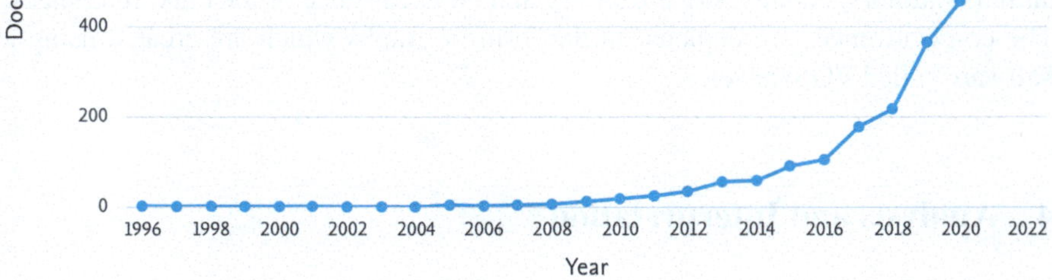

Fig. 1 Research articles published throughout the years

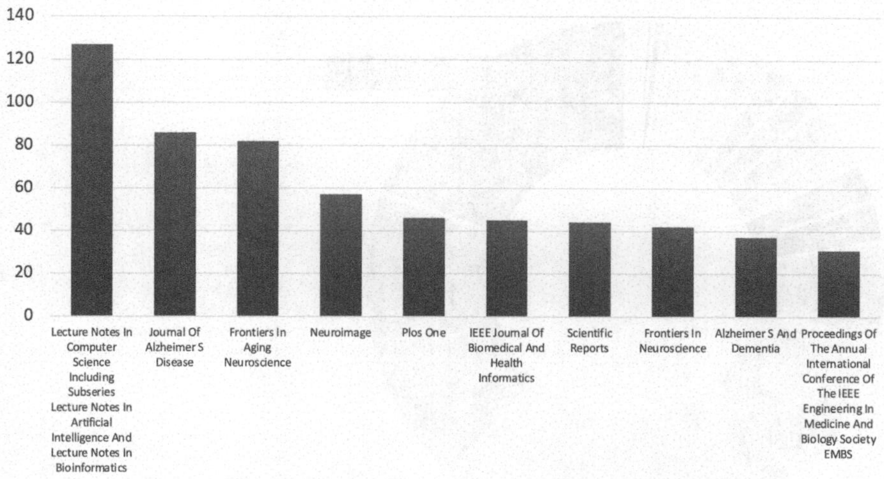

Fig. 2 Top 10 publishing for a globally

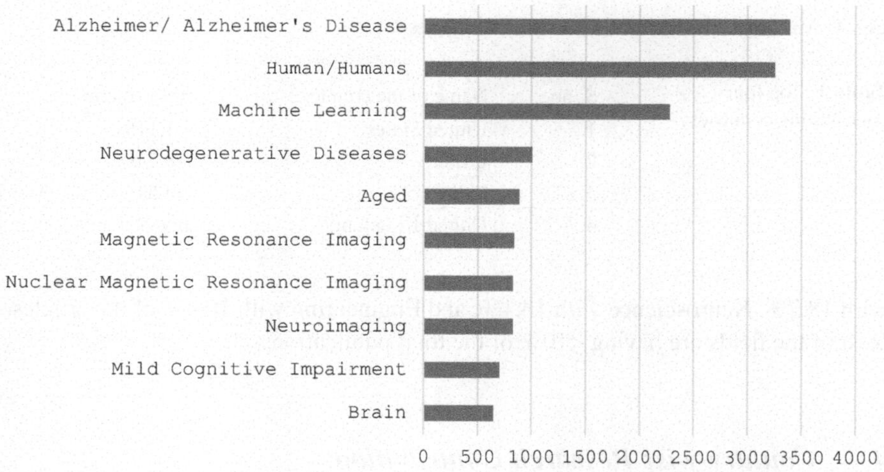

Fig. 3 Most common keywords

4.3 Research Area-wise Distribution of Articles

Since the prediction of Alzheimer's disease using machine learning is helping with the early detection of the disease, it is catching the eyes of researchers from various fields including medicine, computer science, pharmacology, neuroscience, etc. The articles extracted for the analysis in this paper are published in various fields of research. Figure 4 below gives the subject/research area-wise articles distribution. It can be concluded from the distribution that most of the research, i.e., 19.4% of the total publications are done in the field of Computer Science, followed by Medicine

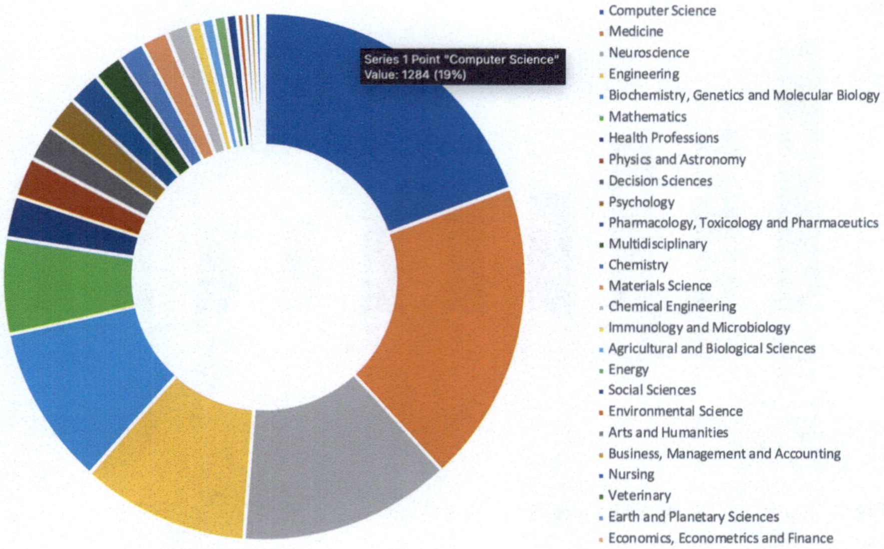

Fig. 4 Area-wise distribution of the research articles

Table 3 Top four contributing countries

S. no.	Name of the country	Contribution
1.	United States	31.11%
2.	China	18.73%
3.	India	13.54%
4.	United Kingdom	10.99%

with 18.7%, Neuroscience with 13.1%, and Engineering with 10.4% of the articles. Rest of the fields are having <10% of the total publications.

4.4 Country-wise Research Contribution

After analyzing the area of research, the country-wise contributions in the field were analyzed. From the extracted articles, it was found that there are around 100 countries who are actively working in the area. However, the publications made by the researchers of the top four countries mentioned in Table 3 below sum up to more than 70% of the total articles published. Thus, it can be concluded that the United States, China, India, and the United Kingdom are way ahead of the other countries.

4.5 Most Citations

The 3182 articles that are considered in this analysis. The total citations of these publications are found to be 55,918. However, the top cited research papers, review papers, and book chapters are given in Tables 4, 5, and 6, respectively.

Some of the authors have contributed to the research remarkably and have citations that are commendable. Figure 5 below shows the authors who have published more than 15 articles.

4.6 Bibliometric Analysis

The bibliometric analysis is a technique in which the large amount of research data is analyzed based on Relationships between publications, authors, countries, citations, etc. In this section, we have conducted the bibliometric analysis by generating co-occurrence graphs based on several criteria.

4.6.1 Co-occurrence of Keywords

Figure 6 below displays the co-occurrence of the keywords in the research articles. The nodes of the graph depict the keyword whereas, the edges show the frequency of its co-occurrence with the other keyword. The size of the nodes depicts the

Table 4 Most cited research papers globally

Reference	Paper title
[33]	"Hierarchical feature representation and multimodal fusion with deep learning for AD/MCI diagnosis"
[34]	"Single subject prediction of brain disorders in neuroimaging: Promises and pitfalls"
[35]	"Multi-modal multi-task learning for joint prediction of multiple regression and classification variables in Alzheimer's disease"
[36]	"Machine learning framework for early MRI-based Alzheimer's conversion prediction in MCI subjects"
[37]	"Linguistic features identify Alzheimer's disease in narrative speech"
[38]	"Multimodal Neuroimaging Feature Learning for Multiclass Diagnosis of Alzheimer's Disease"
[39]	"Predictive markers for AD in a multi-modality framework: An analysis of MCI progression in the ADNI population"
[40]	"Clonally expanded CD8 T cells patrol the cerebrospinal fluid in Alzheimer's disease"
[41]	"Multimodal Neuroimaging Feature Learning with Multimodal Stacked Deep Polynomial Networks for Diagnosis of Alzheimer's Disease"
[42]	"Plasma exosomal miRNAs in persons with and without Alzheimer disease: Altered expression and prospects for biomarkers"

Table 5 Most cited review papers globally

Reference	Paper title
[43]	"Using Support Vector Machine to identify imaging biomarkers of neurological and psychiatric disease: A critical review"
[44]	"Using deep learning to investigate the neuroimaging correlates of psychiatric and neurological disorders: Methods and applications"
[45]	"A review on neuroimaging-based classification studies and associated feature extraction methods for Alzheimer's disease and its prodromal stages"
[46]	"Machine learning in mental health: A scoping review of methods and applications"
[47]	"Random forest algorithm for the classification of neuroimaging data in Alzheimer's disease: A systematic review"
[48]	"A review on a deep learning perspective in brain cancer classification"
[49]	"Recent publications from the Alzheimer's Disease Neuroimaging Initiative: Reviewing progress toward improved AD clinical trials"
[50]	"Molecular docking: Challenges, advances and its use in drug discovery perspective"
[51]	"Applications of machine learning to diagnosis and treatment of neurodegenerative diseases"
[52]	"Multivariate data analysis and machine learning in Alzheimer's disease with a focus on structural magnetic resonance imaging"

Table 6 Most cited book chapters globally

Reference	Title
[53]	"Data science algorithms and techniques for smart healthcare using IoT and big data analytics"
[54]	"Identifying interactions between long noncoding RNAs and diseases based on computational methods"
[55]	"Intrinsically disordered proteins in various hypotheses on the pathogenesis of Alzheimer's and Parkinson's diseases"
[56]	"Use of deep learning for disease detection and diagnosis"
[57]	"Deep learning approaches to electrophysiological multivariate time-series analysis"

occurrence of that keyword. The criterion chosen was to show the keywords that are co-occurring with at least ten other keywords, i.e., the keywords which were not co-occurring with ten other keywords are discarded and are not depicted in the graph. From the given graph, the most frequently used keywords in co-occurrence with other keywords are: "machine learning," "Alzheimer's disease," "mild cognitive impairment," "deep learning," "MRI," "magnetic resonance imaging," and "support vector machine."

4.6.2 Co-authorship Analysis

When a research article is written by more than one author, it is called co-authorship. A co-author is a person who has significantly contributed to the research and is equally responsible for conducting the research and writing the paper. The analysis

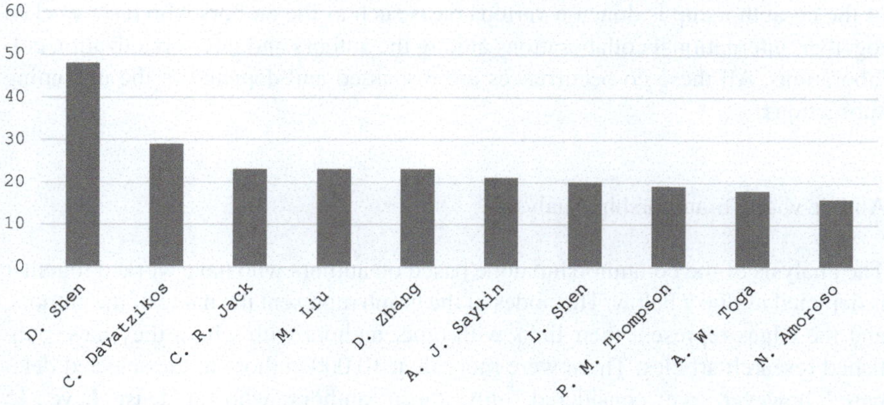

Fig. 5 Authors having more than 15 publications

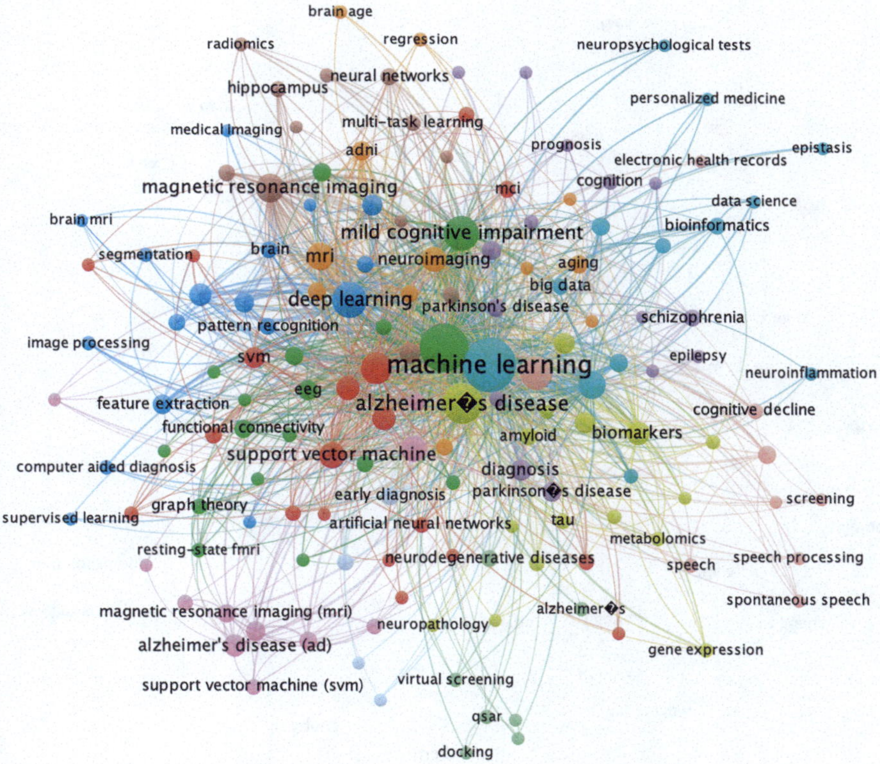

Fig. 6 Co-occurrence graph of keywords

of the co-authorship is done on various basis such as the authors who have worked together, international collaborations among the authors and inter-organization collaborations. All these co-occurrences are discussed and depicted in the upcoming subsections.

Author-wise Co-authorship Analysis

The analysis of the co-authorship done based on authors who have worked together is depicted in Fig. 7 below. The nodes of the graph represent the name of the authors, and the edges represent their links with other authors with whom they have published research articles. There were more than 10,000 authors in the selected database; however, we considered only those authors who at least have 15 publications.

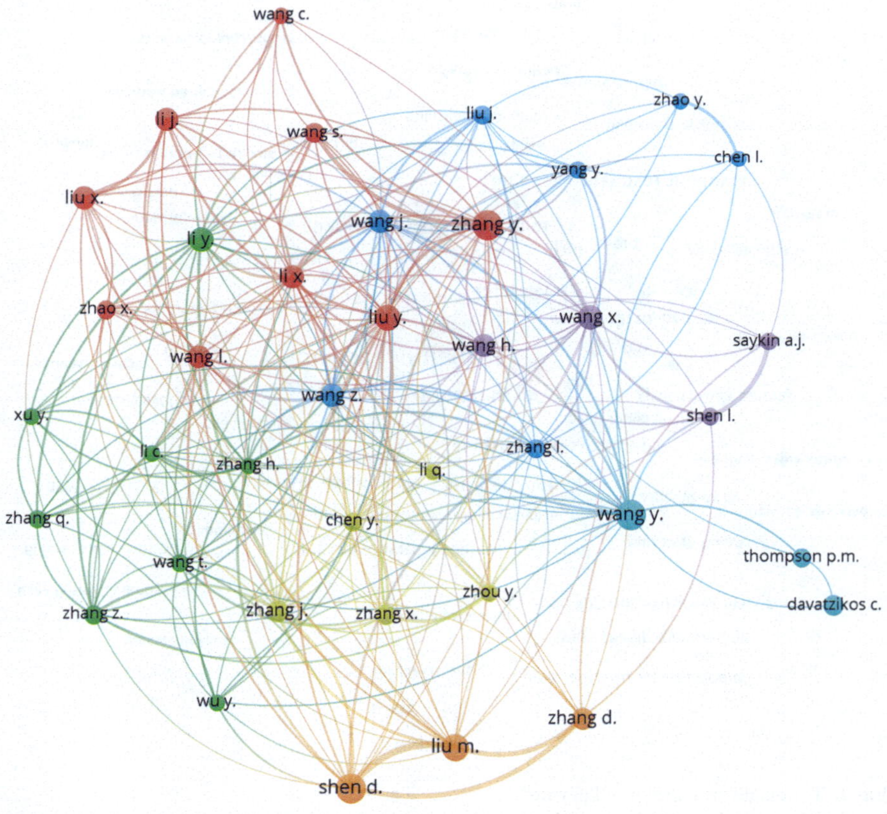

Fig. 7 Co-authorship graph

Country-wise Co-authorship Analysis

As discussed in Sect. 3.4, there are around 100 countries who are working in this research direction. International collaboration depicts that the research area chosen is of high global significance. For analyzing this in detail, a bibliometric graph is generated in which we shortlisted the countries that has at least five publications. Then a graph depicted in Fig. 8 is generated that analyses the co-authorship among the different countries, i.e., the authors from different countries working together on a paper. The nodes represent the name of the countries, whereas the edges represent the authorship links.

Inter-organization Co-authorship Analysis

The analysis of the co-authorship done based on authors belonging to different organizations who have worked together is depicted in Fig. 9 below. The nodes of the graph represent the name of the organizations, and the edges represent their links with other organizations with which they have published research articles. The organizations who have at least published two research articles are considered for the analysis.

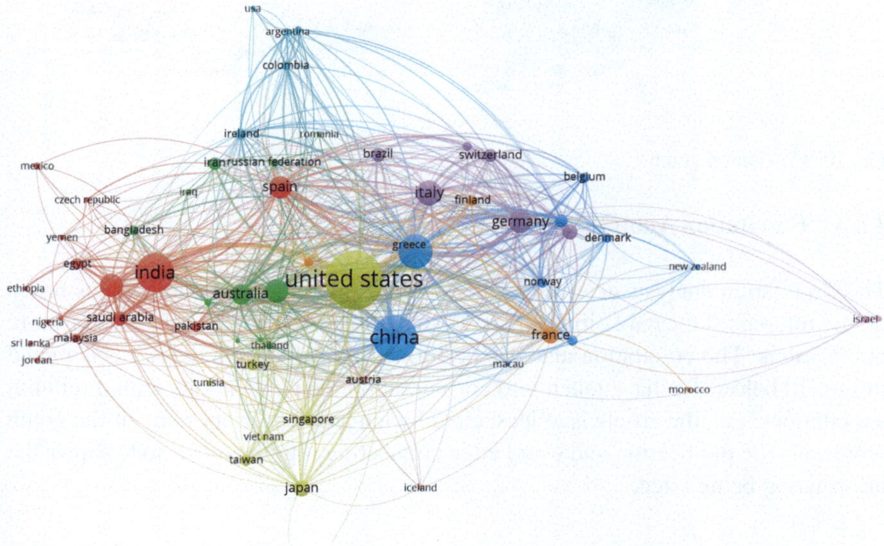

Fig. 8 Country-wise co-authorship graph

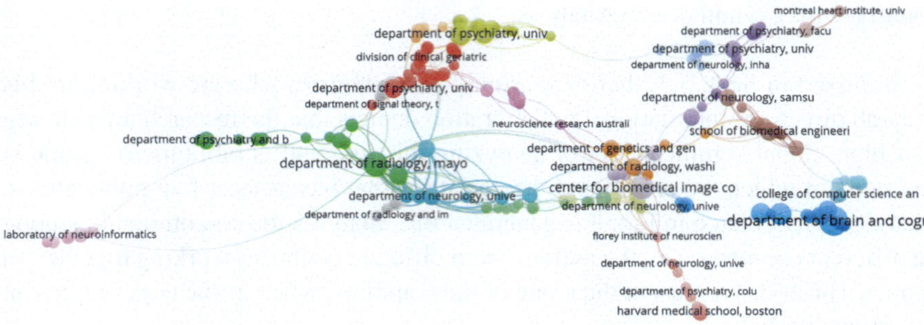

Fig. 9 Inter-organization co-authorship graph

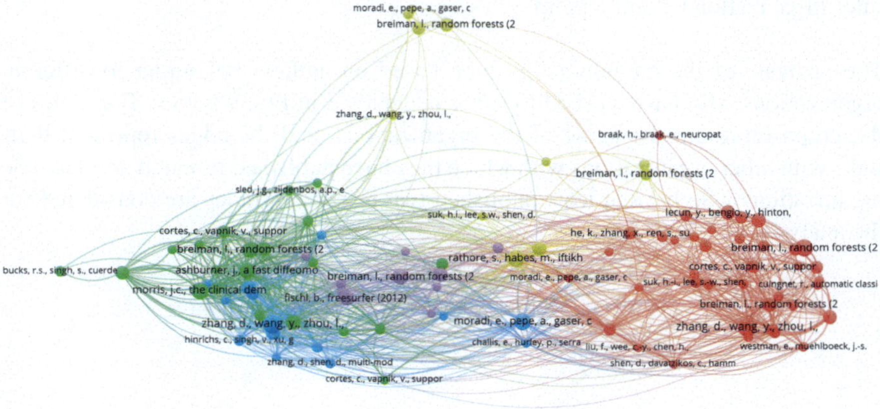

Fig. 10 Co-citation graph

4.6.3 Co-citation Analysis

The Co-citation analysis determines the relationship among the publications based on the references they are citing. It shows the articles that are cited by two or more publications. The co-citation analysis performed on the chosen dataset is depicted in Fig. 10 below. We have taken into consideration the publications with minimum ten citations, i.e., the article is at least cited by ten articles. Every node of the graph represents the publication and every edge connecting it to the other node shows the information being cited.

5 Discussion and Conclusion

Machine learning algorithms could be learned through experience and training. These algorithms have proved their effectiveness in various application domains including healthcare. Talking about the healthcare industry, machine learning techniques can be leveraged for multiple purposes such as monitoring fraudulent insurance claims, medicine supply chain, electronic health record management, and disease prediction. Alzheimer's disease is a brain-related illness that damages the brain tissues and makes the patient unable to perform daily activities. It is commonly found in old people. As per the researchers, this disease is affecting millions of people throughout the globe. So far, no cure has been found for this fatal disease. The symptoms of Alzheimer's are not very noticeable during its initial stage and by the time the symptoms show up it's already very late as the damage to the brain has been done. As per some of the studies conducted in this field, it is suggested that early prediction can help in delaying the damage. Therefore, help can be taken from machine learning algorithms to accurately predict the chances of developing Alzheimer's to prevent it before it's too late.

This article presents an extensive review of the research work done in the field of machine learning to predict or analyze the presence of Alzheimer's disease. We have explored 3182 research articles published in this domain up to the year 2022 to provide the readers with a good understanding of the research statistics. The dataset was extracted from Scopus, with citation information, bibliographical information, abstract, keywords, and references. And a thorough analysis is performed on how many research papers are being published every year, the global fora available for publishing the research articles, the authors, countries and organizations that are actively working in the field, etc.

It can be concluded from the analyses that the number of research articles published per year in this domain is continuously increasing. And a huge surge can be seen in the past 5–6 years, i.e., from 2016 onwards. There are around 150 sources available globally where one can publish or retrieve the articles, the names of the fora with maximum number of publications are also provided in this paper. The keywords that are used most frequently were also analyzed and extracted as "machine learning," "Alzheimer's disease," "mild cognitive impairment," "deep learning," "MRI," "magnetic resonance imaging," etc. The most active field in which the researchers are working on the prediction of Alzheimer's disease using machine learning techniques was found to be "computer science" with 19.4% of the total publications. And the United States was found to be the most active country in the research with 31.1% of the overall publications. We have also provided the details of the publications such as research papers, review papers, book chapters, and the authors that are the most widely cited throughout the globe. A bibliometric analysis is then done on the dataset using VOSviewer to extract various relationships and links among different aspects of the research such as the co-occurrence of keywords, co-authorship based on authors, organizations, and countries and co-citation analysis.

References

1. Simon, A., Deo, M., Selvam, V., & Babu, R. (2016). An overview of machine learning and its applications. *International Journal of Electrical Sciences & Engineering, 1*, 22–24.
2. Alzubi, J., Nayyar, A., & Kumar, A. (2018). Machine learning from theory to algorithms: An overview. *Journal of Physics: Conference Series, 1142*, 012012.
3. Antor, B., Morshedul, A. H. M., Jamil, M., Mamtaz, M. M., Khan, S., Aljahdali, M., Kaur, P. S., & Masud, M. (2021). A comparative analysis of machine learning algorithms to predict Alzheimer's disease. *Journal of Healthcare Engineering, 2021*, 9917919.
4. Masud, M., Eldin Rashed, A. E., & Hossain, M. S. (2022). Convolutional neural network-based models for diagnosis of breast cancer. *Neural Computing and Applications, 34*, 11383.
5. Hsu, C. W., & Lin, C. J. (2002). A comparison of methods for multiclass support vector machines. *IEEE Transactions on Neural Networks, 13*(2), 415–425.
6. Boateng, E. Y., & Abaye, D. A. (2019). A review of the logistic regression model with emphasis on medical research. *Journal of Data Analysis and Information Processing, 7*(4), 190–207.
7. Biau, G. (2012). Analysis of a random forests model. *Journal of Machine Learning Research, 13*, 1063–1095.
8. Aitkenhead, M. J. (2008). A co-evolving decision tree classification method. *Expert Systems with Applications, 34*(1), 18–25.
9. Tayal, A., Gupta, J., Solanki, A., Bisht, K., Nayyar, A., & Masud, M. (2022). DL-CNN-based approach with image processing techniques for diagnosis of retinal diseases. *Multimedia Systems, 28*, 1417.
10. Masud, M., Sikder, N., Nahid, A.-A., Bairagi, A. K., & AlZain, M. A. (2021). A machine learning approach to diagnosing lung and colon cancer using a deep learning-based classification framework. *Sensors, 21*(3), 748.
11. Masud, M., Alhumyani, H., Alshamrani, S. S., Cheikhrouhou, O., Ibrahim, S., Muhammad, G., et al. (2020). Leveraging deep learning techniques for malaria parasite detection using mobile application. *Wireless Communications and Mobile Computing, 2020*, 8895429.
12. Konovalenko, I., & Ludwig, A. (2021). Comparison of machine learning classifiers: A case study of temperature alarms in a pharmaceutical supply chain. *Information Systems, 100*, 101759.
13. Yani, L. P. E., & Aamer, A. (2023). Demand forecasting accuracy in the pharmaceutical supply chain: A machine learning approach. *International Journal of Pharmaceutical and Healthcare Marketing, 17*(1), 1–23.
14. Rukhsar, L., Bangyal, W. H., Nisar, K., & Nisar, S. (2022). Prediction of insurance fraud detection using machine learning algorithms. *Mehran University Research Journal of Engineering & Technology, 41*(1), 33–40.
15. Fiest, K. M., Roberts, J. I., Maxwell, C. J., Hogan, D. B., Smith, E. E., Frolkis, A., et al. (2016). The prevalence and incidence of dementia due to alzheimer's disease: A systematic review and meta-analysis. *The Canadian Journal of Neurological Sciences/Journal Canadien des Sciences Neurologiques, 43*(S1), S51–S82.

16. Rubenstein, E., Hartley, S., & Bishop, L. (2020). Epidemiology of dementia and alzheimer disease in individuals with down syndrome. *JAMA Neurology, 77*(2), 262.
17. Brookmeyer, R., Johnson, E., Ziegler-Graham, K., & Arrighi, H. M. (2007). Forecasting the global burden of Alzheimer's disease. *Alzheimer's and Dementia, 3*(3), 186–191.
18. Reitz, C., Brayne, C., & Mayeux, R. (2011). Epidemiology of alzheimer disease. *Nature Reviews Neurology, 7*(3), 137–152.
19. Sims, I. M. (2009). 2009 Alzheimer's disease facts and figures. *Alzheimer's and Dementia, 5*(3), 234–270.
20. Mendez, M. F., Underwood, K. L., Zander, B. A., Mastri, A. R., Sung, J. H., & Frey, W. H. (1992). Risk factors in Alzheimer's disease: A clinicopathologic study. *Neurology, 42*(4), 770.
21. Etters, L., Goodall, D., & Harrison, B. (2008). Caregiver burden among dementia patient caregivers: A review of the literature. *Journal of the American Academy of Nurse Practitioners, 20*(8), 423–428.
22. Crous-Bou, M., Minguillón, C., Gramunt, N., & Molinuevo, J. L. (2017). Alzheimer's disease prevention: From risk factors to early intervention. *Alzheimer's Research & Therapy, 9*, 71.
23. Vaughn, V. M., & Linder, J. A. (2018). Thoughtless design of the electronic health record drives overuse, but purposeful design can nudge improved patient care. *BMJ Quality and Safety, 27*(8), 583–586.
24. Chi, C.-L., Zeng, W., Oh, W., Borson, S., Lenskaia, T., Shen, X., & Tonellato, P. J. (2017). Personalized long-term prediction of cognitive function: Using sequential assessments to improve model performance. *Journal of Biomedical Informatics, 76*, 78–86.
25. Retrieved from adni.loni.usc.edu
26. Retrieved from aibl.csiro.au
27. Retrieved from www.oasis-brains.org
28. Kazemi, K., & Noorizadeh, N. (2014). Quantitative comparison of SPM, FSL, and brainsuite for brain MR image segmentation. *Journal of Biomedical Physics and Engineering, 4*(1), 13.
29. Dale, A. M., Fischl, B., & Sereno, M. I. (1999). Cortical surface-based analysis: I. Segmentation and surface reconstruction. *Neuroimage, 9*(2), 179–194.
30. Retrieved from https://www.elsevier.com/solutions/scopus/how-scopus-works
31. Retrieved from https://www.elsevier.com/solutions/scopus/how-scopus-works/content
32. Sotirios, K., Johnson, P., Ekstedt, M., & Lagerström, R. (2021). Research communities in cyber security: A comprehensive literature review. *Computer Science Review, 42*, 100431.
33. Suk, H.-I., Lee, S.-W., Shen, D., & Alzheimer's Disease Neuroimaging Initiative. (2014). Hierarchical feature representation and multimodal fusion with deep learning for AD/MCI diagnosis. *NeuroImage, 101*, 569–582.
34. Arbabshirani, M. R., Plis, S., Sui, J., & Calhoun, V. D. (2017). Single subject prediction of brain disorders in neuroimaging: Promises and pitfalls. *NeuroImage, 145*, 137–165.
35. Zhang, D., Shen, D., & Alzheimer's Disease Neuroimaging Initiative. (2012). Multi-modal multi-task learning for joint prediction of multiple regression and classification variables in Alzheimer's disease. *NeuroImage, 59*(2), 895–907.
36. Moradi, E., Pepe, A., Gaser, C., Huttunen, H., Tohka, J., & Alzheimer's Disease Neuroimaging Initiative. (2015). Machine learning framework for early MRI-based Alzheimer's conversion prediction in MCI subjects. *NeuroImage, 104*, 398–412.
37. Fraser, K. C., Meltzer, J. A., & Rudzicz, F. (2016). Linguistic features identify Alzheimer's disease in narrative speech. *Journal of Alzheimer's Disease, 49*(2), 407–422.
38. Liu, S., Sidong, L., Weidong, C., Hangyu, C., Sonia, P., Ron, K., Dagan, F., & Michael, J. F. (2014). Multimodal neuroimaging feature learning for multiclass diagnosis of Alzheimer's disease. *IEEE Transactions on Biomedical Engineering, 62*(4), 1132–1140.
39. Hinrichs, C., Singh, V., Xu, G., Johnson, S. C., & Alzheimers Disease Neuroimaging Initiative. (2011). Predictive markers for AD in a multi-modality framework: An analysis of MCI progression in the ADNI population. *NeuroImage, 55*(2), 574–589.

40. Gate, D., Saligrama, N., Leventhal, O., Yang, A. C., Unger, M. S., Middeldorp, J., et al. (2020). Clonally expanded CD8 T cells patrol the cerebrospinal fluid in Alzheimer's disease. *Nature, 577*(7790), 399–404.
41. Shi, J., Zheng, X., Li, Y., Zhang, Q., & Ying, S. (2017). Multimodal neuroimaging feature learning with multimodal stacked deep polynomial networks for diagnosis of Alzheimer's disease. *IEEE Journal of Biomedical and Health Informatics, 22*(1), 173–183.
42. Lugli, G., Cohen, A. M., Bennett, D. A., Shah, R. C., Fields, C. J., Hernandez, A. G., & Smalheiser, N. R. (2015). Plasma exosomal miRNAs in persons with and without Alzheimer disease: Altered expression and prospects for biomarkers. *PLoS One, 10*(10), e0139233.
43. Orru, G., Pettersson-Yeo, W., Marquand, A. F., Sartori, G., & Mechelli, A. (2012). Using support vector machine to identify imaging biomarkers of neurological and psychiatric disease: A critical review. *Neuroscience & Biobehavioral Reviews, 36*(4), 1140–1152.
44. Vieira, S., Pinaya, W. H. L., & Mechelli, A. (2017). Using deep learning to investigate the neuroimaging correlates of psychiatric and neurological disorders: Methods and applications. *Neuroscience & Biobehavioral Reviews, 74*, 58–75.
45. Rathore, S., Habes, M., Iftikhar, M. A., Shacklett, A., & Davatzikos, C. (2017). A review on neuroimaging-based classification studies and associated feature extraction methods for Alzheimer's disease and its prodromal stages. *NeuroImage, 155*, 530–548.
46. Shatte, A. B. R., Hutchinson, D. M., & Teague, S. J. (2019). Machine learning in mental health: A scoping review of methods and applications. *Psychological Medicine, 49*(9), 1426–1448.
47. Sarica, A., Cerasa, A., & Quattrone, A. (2017). Random forest algorithm for the classification of neuroimaging data in Alzheimer's disease: A systematic review. *Frontiers in Aging Neuroscience, 9*, 329.
48. Tandel, G. S., Biswas, M., Kakde, O. G., Tiwari, A., Suri, H. S., Turk, M., et al. (2019). A review on a deep learning perspective in brain cancer classification. *Cancers, 11*(1), 111.
49. Weiner, M. W., Veitch, D. P., Paul, S. A., Laurel, A. B., Cairns, N. J., Green, R. C., et al. (2017). Recent publications from the Alzheimer's Disease Neuroimaging Initiative: Reviewing progress toward improved AD clinical trials. *Alzheimer's & Dementia, 13*(4), e1–e85.
50. Saikia, S., & Bordoloi, M. (2019). Molecular docking: Challenges, advances and its use in drug discovery perspective. *Current Drug Targets, 20*(5), 501–521.
51. Myszczynska, M. A., Poojitha, N. O., Lacoste, M. B. A., Neil, D., Saffari, A., Mead, R., et al. (2020). Applications of machine learning to diagnosis and treatment of neurodegenerative diseases. *Nature Reviews Neurology, 16*(8), 440–456.
52. Falahati, F., Westman, E., & Simmons, A. (2014). Multivariate data analysis and machine learning in Alzheimer's disease with a focus on structural magnetic resonance imaging. *Journal of Alzheimer's Disease, 41*(3), 685–708.
53. Syed, L., Jabeen, S., Manimala, S., & Elsayed, H. A. (2019). Data science algorithms and techniques for smart healthcare using IoT and big data analytics. In *Smart techniques for a smarter planet: Towards smarter algorithms* (pp. 211–241). Springer.
54. Lan, W., Huang, L., Lai, D., & Chen, Q. (2018). Identifying interactions between long noncoding RNAs and diseases based on computational methods. In *Computational systems biology: Methods and protocols* (pp. 205–221). Springer.
55. Coskuner, O., & Uversky, V. N. (2019). Intrinsically disordered proteins in various hypotheses on the pathogenesis of Alzheimer's and Parkinson's diseases. *Progress in Molecular Biology and Translational Science, 166*, 145–223.
56. Mishra, S., Dash, A., & Jena, L. (2021). Use of deep learning for disease detection and diagnosis. In *Bio-inspired neurocomputing* (pp. 181–201). Springer.
57. Morabito, F. C., Campolo, M., Ieracitano, C., & Mammone, N. (2019). Deep learning approaches to electrophysiological multivariate time-series analysis. In *Artificial intelligence in the age of neural networks and brain computing* (pp. 219–243). Academic Press.
58. Ellis, K. A., Bush, A. I., Darby, D., de Fazio, D., Foster, J., Hudson, P., et al. (2009). The Australian Imaging, Biomarkers and Lifestyle (AIBL) study of aging: Methodology and base-

line characteristics of 1112 individuals recruited for a longitudinal study of Alzheimer's disease. *International Psychogeriatrics, 21*, 672. https://doi.org/10.1017/S1041610209009405
59. Zhang, J., Xiao, M., Gao, L., & Chu, S. (2019). Probability and interval hybrid reliability analysis based on adaptive local approximation of projection outlines using support vector machine. *Computer-Aided Civil and Infrastructure Engineering, 34*(11), 991–1009.
60. Rabeh, A. B., Benzart, F. I., & Amiri, F. (2016). Diagnosis of alzheimer diseases in early step using SVM (support vector machine). In *2016 13th International Conference on Computer Graphics, Imaging and Visualization (CGiV), Beni Mellal* (pp. 364–367). IEEE. https://doi.org/10.1109/CGiV.2016.76
61. Khedher, L., Illan, I. A., Gorriz, J. M., Ramirez, J., Brahim, A., & Meyer-Baese, A. (2017). Independent component analysis-support vector machine-based computer-aided diagnosis system for Alzheimer's with visual support. *International Journal of Neural Systems, 27*(3), 1650050.
62. Zeng, N., Qiu, H., Wang, Z., Liu, W., Zhang, H., & Li, Y. (2018). A new switching-delayed-PSO-based optimized SVM algorithm for diagnosis of Alzheimer's disease. *Neurocomputing, 320*, 195–202.
63. Bi, X., Shu, Q., Sun, Q., & Xu, Q. (2018). Random support vector machine cluster analysis of resting-state of MRI in Alzheimer's disease. *PLoS One, 13*(3), e0194479. https://doi.org/10.1371/journal.pone.0194479
64. Ferreira, L. K., Rondina, J. M., Kubo, R., Ono, C. R., Leite, C. C., Smid, J., et al. (2017). Support vector machine-based classification of neuroimages in Alzheimer's disease: direct comparison of FDG-PET, rCBF-SPECT and MRI data acquired from the same individuals. *Brazilian Journal of Psychiatry, 40*(2), 181–191. https://doi.org/10.1590/1516-4446-2016-2083
65. Gupta, Y., Lama, R., Kwon, G. R., & Alzheimer's Disease Neuroimaging Initiative. (2019). Prediction and classification of Alzheimer's disease based on combined features from apolipoprotein-E genotype, cerebrospinal fluid, MR, and FDG-PET imaging biomarkers. *Frontiers in Computational Neuroscience, 13*, 72. https://doi.org/10.3389/fncom.2019.00072
66. Alickovic, E., & Subasi, A. (2019). Automatic detection of Alzheimer disease based on histogram and random forest. In *International Conference on medical imaging and biological engineering, CMBEBIH 2019. CMBEBIH 2019. IFMBE Proceedings* (Vol. 73). Springer. https://doi.org/10.1007/978-3-030-17971-7_14
67. Bloch, L., & Friedrich, C. M. (2019). Classification of Alzheimer's Disease using volumetric features of multiple MRI scans. In *2019 41st Annual International Conference of the IEEE Engineering in Medicine and Biology Society (EMBC), Berlin, Germany* (pp. 2396–2401). IEEE. https://doi.org/10.1109/EMBC.2019.8857188
68. Lu, S., Xia, Y., Cai, W., Fulham, M., & Feng, D. D. (2017). Early identification of mild cognitive impairment using incomplete random forest-robust support vector machine and FDG-PET imaging. *Computerized Medical Imaging and Graphics, 60*, 35–41.
69. Moore, P. J., Lyons, T. J., Gallacher, J., & Ginsberg, S. D. (2019). Random forest prediction of Alzheimer's disease using pairwise selection from time series data. *PLoS One, 14*(2), e0211558. https://doi.org/10.1371/journal.pone.0211558
70. Silva, I. R. R., Silva, G. S. L., de Souza, R. G., dos Santos, W. P., & de A. Fagundes, R. A. (2019). Model based on deep feature extraction for diagnosis of Alzheimer's disease. In *2019 International Joint Conference on Neural Networks (IJCNN), Budapest, Hungary* (pp. 1–7). IEEE. https://doi.org/10.1109/IJCNN.2019.8852138
71. Leming, M., Gorriz, J. M., & Suckling, J. (2020). Ensemble deep learning on large, mixed-site fMRI datasets in autism and other tasks. *International Journal of Neural Systems, 30*(7), 2050012.
72. Xu, J., Gui, C., & Han, Q. (2020). Recognition of rust grade and rust ratio of steel structures based on ensembled convolutional neural network. *Computer-Aided Civil and Infrastructure Engineering, 35*(10), 1160–1174.

73. Vera-Olmos, F. J., Pardo, E., Melero, H., & Malpica, N. (2019). DeepEye: Deep convolutional network for pupil detection in real environments. *Integrated Computer-Aided Engineering, 26*(1), 85–95.
74. Zhang, A., Wang, K. C. P., Fei, Y., Liu, Y., Chen, C., Yang, G., et al. (2019). Automated pixel-level pavement crack detection on 3D asphalt surfaces with a recurrent neural network. *Computer-Aided Civil and Infrastructure Engineering, 34*(3), 213–229.
75. An, N., Ding, H., Yang, J., Au, R., & Ang, T. F. A. (2020). Deep ensemble learning for Alzheimer's disease classification. *Journal of Biomedical Informatics, 105*, 103411. https://doi.org/10.1016/j.jbi.2020.103411
76. Ebrahimighahnavieh, M. A., Luo, S., & Chiong, R. (2020). Deep learning to detect Alzheimer's disease from neuroimaging: A systematic literature review. *Computer Methods and Programs in Biomedicine, 187*, 105242. https://doi.org/10.1016/j.cmpb.2019.105242
77. Jo, T., Mho, K., & Saykin, A. (2019). Deep learning in Alzheimer's disease: Diagnostic classification and prognostic prediction using neuroimaging data. *Frontiers in Aging Neuroscience, 11*, 220. https://doi.org/10.3389/fnagi.2019.00220
78. Lee, G., Nho, K., Kang, B., Sohn, K. A., Kim, D., & for Alzheimer's Disease Neuroimaging Initiative. (2019). Predicting Alzheimer's disease progression using multi-modal deep learning approach. *Scientific Reports, 9*(1), 1952. https://doi.org/10.1038/s41598-018-37769-z
79. Wada, A., Tsuruta, K., Irie, R., Kamagata, K., Maekawa, T., Fujita, S., et al. (2019). Differentiating Alzheimer's disease from dementia with lewy bodies using a deep learning technique based on structural brain connectivity. *Magnetic Resonance in Medical Sciences, 18*(3), 219–224. https://doi.org/10.2463/mrms.mp.2018-0091
80. Rajesh Kumar, P., Arun Prasath, T., Pallikonda Rajasekaran, M., & Vishnuvarthanan, G. (2018). Brain subject estimation using PSO k-means clustering - An automated aid for the assessment of clinical dementia. In S. Satapathy & A. Joshi (Eds.), *Information and communication technology for intelligent systems (ICTIS 2017) - Volume 1. ICTIS 2017. Smart innovation, systems and technologies* (Vol. 83). Springer. https://doi.org/10.1007/978-3-319-63673-3_58
81. Li, F., Liu, M., & The Alzheimer's Disease Neuroimaging Initiative. (2018). Alzheimer's disease diagnosis based on multiple cluster dense convolutional networks. *Computerized Medical Imaging and Graphics, 70*, 101–110.
82. Farouk, Y., & Rady, S. (2020). Early diagnosis of Alzheimer's disease using unsupervised clustering. *International Journal of Intelligent Computing and Information Sciences, 20*(2), 112–124.
83. Alexander, N., Alexander, D. C., Barkhof, F., & Denaxas, S. (2020). Using unsupervised learning to identify clinical subtypes of Alzheimer's disease in electronic health records. *Studies in Health Technology and Informatics, 270*, 499–503. https://doi.org/10.3233/SHTI200210
84. Ortiz-Garcia, A., Munilla, J., Gorriz, J. M., & Ramirez, J. (2016). Ensembles of deep learning architectures for the early diagnosis of Alzheimer's disease. *International Journal of Neural Systems, 26*(7), 1650025.
85. Martinez-Murcia, F. J., Górriz, J. M., Ramírez, J., & Ortiz, A. (2016). A structural parametrization of the brain using hidden markov models-based paths in Alzheimer's disease. *International Journal of Neural Systems, 26*(07), 1650024. https://doi.org/10.1142/S0129065716500246
86. Jain, R., Jain, N., Aggarwal, A., & Hemanth, D. J. (2019). Convolutional neural network based Alzheimer's disease classification from magnetic resonance brain images. *Cognitive Systems Research, 57*, 147–159.
87. Farooq, A., Anwar, S., Awais, M., & Rehman, S. (2017). A deep CNN based multi-class classification of Alzheimer's disease using MRI. In *2017 IEEE International Conference on Imaging Systems and Techniques (IST), Beijing* (pp. 1–6). IEEE. https://doi.org/10.1109/IST.2017.8261460
88. Idrees, S. M., Alam, M. A., Agarwal, P., & Ansari, L. (2019). Effective predictive analytics and modeling based on historical data. In *Advances in computing and data sciences: Third International Conference, ICACDS 2019, Ghaziabad, India, April 12–13, 2019. Revised selected papers, part II* (pp. 552–564). Springer Singapore.

89. Saiyeda, A., Idrees, S. M., & Alam, M. A. (2019). Optimizing learning using student activity tracking data with machine learning techniques. In *Advances in computing and data sciences, ICIDSSD '19* (p. 33). Springer.
90. Ramana, K., Kumar, M. R., Sreenivasulu, K., Gadekallu, T. R., Bhatia, S., Agarwal, P., & Idrees, S. M. (2022). Early prediction of lung cancers using deep saliency capsule and pre-trained deep learning frameworks. *Frontiers in Oncology, 12*, 886739.
91. Xu, L., Liang, G., Liao, C., Chen, G., & Chang, C. (2018). An efficient classifier for Alzheimer's disease genes identification. *Molecules, 23*(12), 3140.

Applicability and Integration of Blockchain in Healthcare

Tawseef Ahmed Teli, Sheikh Umar Mushtaq, and Waseem Ahmad Bhat

1 Introduction

Blockchain is a trending technology which works on the principles of immutability and consensus. This technology complements the need for privacy, security, and information exchange in a healthcare environment. Any medication given to a patient, the sequence of diagnosis and the possible prognosis all sum up as transactions on the blockchain, which makes health professionals accountable. The management of data in a healthcare system is a nuisance due to the myriad stakeholders involved and the sensitivity of the data. However, the integration of blockchain enhances the management of this data with properties that are desired in various aspects of the healthcare system. Figure 1 provides an overview of blockchain in healthcare.

1.1 The Imperative for a Robust Healthcare System

Despite unprecedented health achievements in the last centuries, the total number of annual deaths globally will grow in the decades ahead due to population growth, population aging, and persevering disparities. The COVID-19 pandemic has further

T. A. Teli (✉)
Higher Education Department, Jammu, Jammu and Kashmir, India

S. U. Mushtaq
Lovely Professional University, Phagwara, Punjab, India

W. A. Bhat
North Campus, University of Kashmir, Srinagar, Jammu and Kashmir, India
e-mail: wab.cs@uok.edu.in

S. M. Idrees et al. (eds.), *Blockchain and Machine Learning Innovations*, Signals and Communication Technology,
https://doi.org/10.1007/978-3-031-88095-7_10

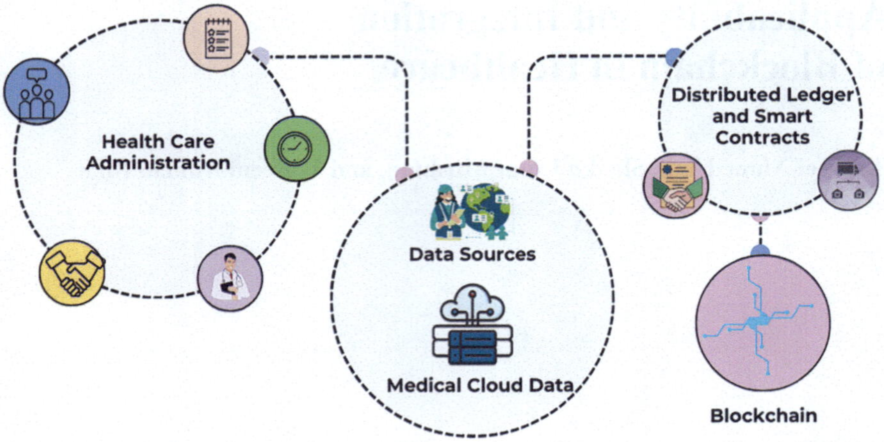

Fig. 1 Blockchain in healthcare

underscored the crucial need to address these encounters, revealing serious gaps in community health systems and exposing the overwhelming consequences of discriminations in healthcare access and excellence. It is imperious to double down on efforts to go beyond typical public health dealings. Current indication shows that healthcare is often insufficient and of deprived quality across countries, with the maximum vulnerable populations proceeding to the worst. The United Nations Agenda 2030 set Universal Health Coverage (UHC) as a sustainable development target, emphasizing the need for high-quality healthcare systems [1]. The lessons from the COVID-19 pandemic highlight the significance of resilience and flexibility in healthcare systems, underscoring the value of improvement initiatives in times of crisis [2]. This transformative change in health policy project and implementation places quality of care at the heart of the international health program. Health improvements can remain indefinable unless services are of adequate quality to be effective. The transformation of healthcare will necessitate an extraordinary commitment to enhance the quality, drawing on the practices of COVID-19 to strengthen readiness, equity, and competence [3]. With the global momentum of UHC as a background and the insights gained from managing a worldwide health crisis, the time is suitable for advancing superior healthcare system in a given context. This can be done by consistently bringing care that recovers or upholds health outcomes. Quality of care must become a political priority. Policies need to change by prioritizing UHC, investing in excellent schemes, engaging in constant learning and improvement, implementing tracer indicators to reorient health schemes toward outcomes that matter to people. The COVID-19 pandemic has established that high-quality care is not just an ambition but obligation to defense global health against future threats.

1.2 The Complex Healthcare Ecosystem

Handling complexity within healthcare ecosystems signifies a mounting universal challenge that requires a multidimensional and collaborative approach. This challenge is predominantly pronounced in light of recent calls for improved care integration to improve the levels of comorbidity and frailty within populations [4]. Key areas in this procedure include governance outlines, resource allocation, service organization, and risk supervision. Each domain demands nuanced thought and careful adjustment to balance the stresses of integrated care distribution for convenience welfare and sustainability [5]. The ecosystem framework for the healthcare service segment is designed to cover a broad range of stakeholders, including essential health providers such as hospitals and pharmaceutical producers and complementary like diagnostic labs, drugstores, and start-ups. It further participates in homecare providers, social media stages, R&D laboratories, software developers, medical colleges, yoga and gym centers, and others [6]. This varied network influences digital technologies to progress into a highly associated and collaborative ecosystem. The need of the hour in a healthcare system is to provide timely and patient care. The integration of new technologies like blockchain, machine learning, and many more will help in strengthening the weaknesses that are currently persistent in healthcare. These weaknesses or limitations inhibit achieving the best healthcare possible and hamper collaborative work for efficient and optimal outcomes. In a healthcare system, there are multiple stakeholders, including healthcare professionals, insurers, and other departments to have a multi-layered approach to overcoming the present limitations, all stakeholders must collaborate in terms of policy, making patient care as well as adaptability [7] toward an efficient healthcare system. All this can only be achieved by integrating many new technologies, blockchain being the most promising one.

1.3 Current Challenges in Healthcare

Significantly, the efficiency of a hospital is evaluated to provide an optimized resource utilization and allocation to reduce the costs incurred and enhance efficiency in a healthcare system [8]. In the same way, the hospital resources must be used optimally. It is also mandatory that the inefficiencies and gaps may be identified, and strategies may be advised to eliminate the inefficiencies [9]. Various works have been done to recognize these inefficiencies as mentioned as provided in the WHO report [10]. In this report, the authors focused on how resources were not allocated appropriately in the disruptions in workflow and the inefficient use of trending technologies like blockchain [11] and machine learning within the healthcare system. Other areas include evaluating the security of patient data using blockchain, the security of medical devices and the security of hospital operations against cyber threats in general. It is imperative that these issues are discussed in detail to provide comprehensive solutions to secure patient information as well as provide efficient healthcare. Resource utilization may come under a strategy that will focus on reducing the monetary expenditure, but more focus shall be assigned to safeguarding patient information and securing medical devices. In its current form, healthcare systems lack standardization, and the sharing of information across various platforms and organizations is challenging. Data silos result in repeated and duplicated efforts and administrative burdens, increasing operational costs and reducing efficiency in resource utilization [12]. Data silos in healthcare refer to the fragmented storage of health statistics among different systems or organizations, generating barriers to data sharing and integration. This fragmentation poses significant challenges to the efficiency, quality, and coordination of healthcare delivery.

2 The Promise of Blockchain

Storing data is only the first step; secure and reliable sharing is the next challenge. This is where blockchain becomes relevant. Its consensus mechanisms ensure data legitimacy and transaction security. By prioritizing patient-centricity and enhancing data privacy and interoperability, the blockchain has the potential to revolutionize healthcare management [13]. Currently, these technologies play a principal role in the healthcare area. ML in healthcare is primarily utilized for executive tasks, medical decision provision, and providing recommendations for analysis and treatment. However, blockchain ensures secure administration of the drug stock chain and protects patient health records. It also enhances security for medical service areas like insurance and others. ML analyzes large volumes of data, forecasts outcomes, provides risk ratings and helps in allocating medical resources precisely. The analysis of implementing various machine learning methods is discussed in [14]. This includes different ML techniques like Random Forest Algorithm, SVM (Support Vector Machine), Decision Trees, and other deep learning techniques. These emerging technologies act as influential tools for clinicians for refining the healthcare schemes. The findings presented in [15] highlight that ML and Blockchain technology can be operably and efficiently integrated in the healthcare field. These findings provide state-of-the-art suggestions on merging of these two transformative technologies and establish their potential application in advancing medical care. Leveraging the progressions in technology, the innovative solutions are proposed in [16] to address healthcare management questions through the integration of blockchain. It is well-suited for ensuring the reliability and safekeeping of healthcare data through consensus mechanisms.

2.1 *Security and Transparency*

Blockchain has equipped the technology with the transformative strength particularly in healthcare, addressing serious challenges in data security, reliability, and transparency. Its decentralized and absolute nature proposes a healthy framework for handling sensitive healthcare data. The cryptographic algorithms and consensus nature of blockchain confirms data integrity and crafts trust between stakeholders.

2.2 *Fundamentals of Blockchain*

In this day and age of Big Data, the healthcare system is facing an important challenge of catering to providing a secure and manageable way of storing and exchanging patient data. This is one of the problems that may be solved by blockchain. It's a technology that works as a distributor ledger which stores transactions as blocks which are unmutable [17]. A blockchain is a technology that inherently provides integrity and security due to its inherent properties. The hash mechanism of blockchain provides a temper-resistant tool concerning records saved on the blockchain, transparency, and accountability in the whole healthcare system. Since blockchain works with consensus strategies, it provides security as well as reliability of healthcare information effectively and efficiently.

2.3 *Blockchain Applications in Key Healthcare Domains*

Blockchain technology has played a vital role in transforming healthcare by providing extra security and privacy of patient information. This technology, when integrated with the pharmaceutical supply chain, provides secure mechanisms in terms of counterfeit drugs and traceability. It also helps in providing clinical trial transparency and assistance in smooth insurance claims through smart contracts. The pharmaceutical supply chain is considered as a critical constituent of universal healthcare organizations, responsible for ensuring drug availability and safety. However, the industry encounters significant challenges including fake/Counterfeit drugs and a deficiency of traceability. These challenges may threaten public health and erode trust in healthcare systems if not addressed.

Counterfeiting can cause severe consequences, as they frequently lack active ingredients and contain inappropriate dosages with harmful substances. Counterfeit medicines have captured the universal markets over the period of time, affecting and risks large lives [18]. Solutions to address them have become inevitable. Therefore, applying effective solutions to address these challenges has become inevitable to enable the integrity and efficacy of the pharmaceutical supply chain. Attending this issue involves employing active verification procedures during all phases of the supply chain. Here, blockchain plays a vital role by creating absolute transactions and allocating a unique identifier to every batch. The system is proposed in [19] to leverage blockchain which lets stakeholders to validate authenticity at all phases of the supply chain. Moreover, advanced cryptographic procedures and smart conventions can ensure that only authenticated entities contribute to the supply chain.

Traceability is another important entity for following the source and handling pharmaceutical drugs during the supply chain. Blockchain can ensure the monitoring of drug journey by providing end-to-end visibility among stakeholders. This combination approves promise with storage circumstances and aids to detect and avoid potential holes in supply chain. The associations among pharmaceutical corporations, regulatory experts, and technology service providers are important to influence these solutions globally. In [20] difficulties in pharmaceutical supply chains, such as drug counterfeiting, and the chances of using digital technologies like blockchain to overcome them were explored. It also assesses the risks connected with the use of these advanced technologies via expert meetings. Additionally, in [21] the system architecture was proposed with detailed algorithms to control working standards for improving traceability within drug supply chain. In this chain, several authors in [22] propose the system that allows the patients to securely access their health record electronically. The data is traceable throughout the system. The healthcare engineering generates immense amount of data on regular basis. Supporting this, RBC Capital Markets claimed in 2019 that the healthcare industry generates 30% of global data approximately [23]. Additionally, they claimed that 50 petabytes of data are produced per year by only hospitals which include health images, EHRs, sensor readings, and other types of tests in the laboratory [24].

The supervision and interchange of Electronic Health Records (EHRs) presents noteworthy challenges inside the healthcare segment. Patchy systems with inconsistent principles and constant security concerns inhibit unified interoperability. These restrictions not only affect the efficient distribution of critical patient statistics but also unfavorably affect the value of patient healthcare and the overall efficiency of healthcare organizations. In this regard, novel resolutions are proposed in [25] to enhance data integrity while maintaining security and transparency in EHRs. EHR interoperability is also considered by introducing a DLT-based solution which provides a novel way to share medicinal data securely. Various essential protocols were identified by integrating existing healthcare mechanisms and DLT for easing smooth transitions and improved interoperability.

Blockchain technology has also a strong potential to transform insurance claims particularly in the healthcare system. This can be done by handling automated verification, enhancing transparency, and dropping fraud. The cumulative density of insurance claims frequently encourages the occurrences of fraud, necessitate the enlargement of that can streamline the processes while improving the discovery of fraud detection. The study in [26] suggests a blockchain-based structure integrated with advanced Artificial Intelligence and machine learning techniques to address these necessities. By implementing smart agreements, automatic procedures can be implemented for processing and validating claims. Transactions are recorded securely

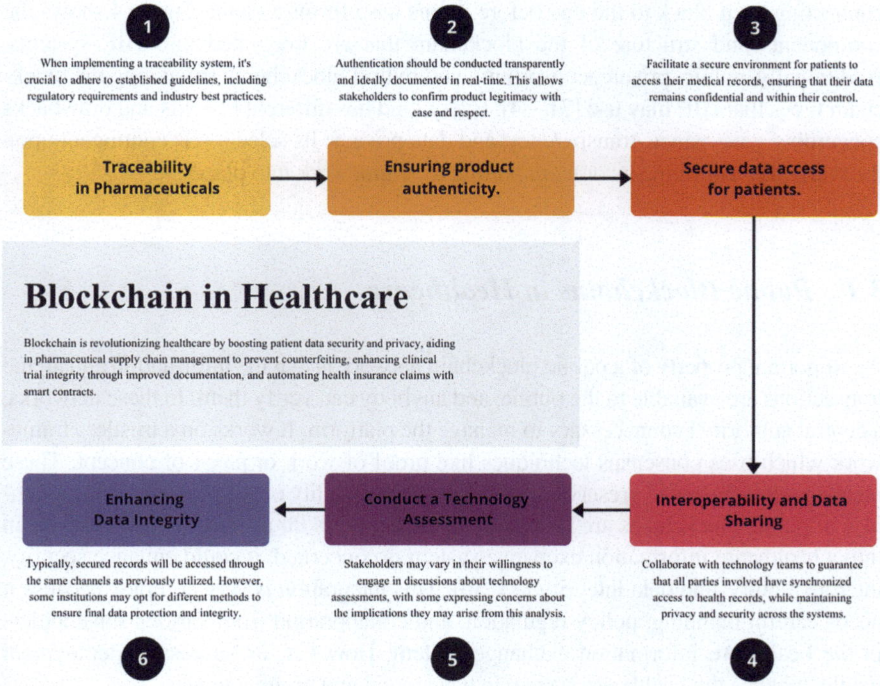

Fig. 2 Blockchain applicability in healthcare

by employing immutable ledger technology of blockchain and thereby reducing duplicitous activities by preventing interference. Further, the transparent nature of blockchain benefits both insurers and policyholders by offering and accelerating the streamlined payments. This transparent tactic of blockchain raises efficiency and trust in the management of healthcare insurance. The summary is given in Fig. 2.

A key component of contemporary healthcare is healthcare information exchange. Hospitals, clinics, laboratories, pharmacies, and other providers may all share health information electronically thanks to it [27–30]. Healthcare practitioners may access vital data in real-time thanks to these solutions, which provide the safe and easy movement of patient data between platforms. It significantly raises the standard of patient care, less medical mistakes, and boosts the effectiveness of healthcare. It provides professionals with a comprehensive picture of a patient's history to focus on targeted treatment regimens. All healthcare practitioners may make well-informed judgments about patient care thanks to this holistic viewpoint.

3 Combining Blockchain in Healthcare

A blockchain is essentially a linked list of blocks, with each block containing transaction data. These transactions are progressively compiled and appended to the blockchain. The data in each block is also used to generate a hash, which is a unique identifier. By connecting each block to the one before it, this hash forms a chain. Figure 4 shows the components and structure of the blockchain that are integrated into HIE systems. Additionally, public, private, consortium, and hybrid blockchains are among the blockchain types that HIE may use [31–34]. Every kind has different benefits and drawbacks concerning governance, transparency, and data privacy. Its selection is contingent upon the particular requirements or team putting the framework into place.

3.1 Public Blockchains in Healthcare

An important property of a public blockchain network is that the information and all the transactions are available to the public, and anybody can verify them. In these networks, a central authority is unnecessary to manage the platform. It works on a trustless framework which uses consensus techniques like proof of work or proof of concept. These mechanisms are used to preserve the security and reliability of the network. Some examples of public blockchains are Ethereum and Bitcoin. As far as integrating a blockchain into a healthcare information exchange system is concerned, it could enhance security, interoperability, and data integrity [35, 36]. This integration is very complex because it needs careful planning, policy regulation, and collaboration from various stakeholders in the healthcare information exchange system. However, its successful deployment would enhance the healthcare system in a profound and promising way.

3.2 Private Blockchains in Healthcare

In a private blockchain network, access is granted only to a restricted number of members. This network is also known as a permissioned network. Only a particular organization or a group of people control the permissions and governance of the private blockchain as compared to the decentralized public blockchain. The usage of a private blockchain network requires proper authentication and authorization. The advantages include enhanced security, optimal interoperability, and transparency [36–38]. Integrating this kind of blockchain technology in a healthcare system is a very complex task of complex designs, planning and efficient execution. Private blockchain integration in the healthcare system would enhance data exchange and patient-centric care by creating smart contracts while following proper policy compliance.

3.3 Consortium Blockchains in Healthcare

In this type of blockchain network, several entities run collaboratively. There is cooperation among several parties, but these blockchains have limited permission access compared to public blockchain networks [39]. It is pertinent to mention that only those members who have been provided permission to join and confirm transactions and actively part speed in consensus building. This type of blockchain is inherently fit to work in an environment in which several stakeholders work together and exchange information while preserving anonymity. There are several advantages of integrating a consortium blockchain in a healthcare system. However, the complexity of integration remains the same as was seen in public and private blockchain networks.

3.4 *Hybrid Blockchains in Healthcare*

A hybrid blockchain is a combination of public and private blockchain networks in which the advantages of both blockchains are retained while reducing any disadvantages. In this type of network, privacy features from the private blockchain network are extensively used, while the transparency, decentralized control, and openness of transactions from the public blockchain network are encouraged. Hyperledger Fabric and Corda are two of the hybrid blockchains that have surfaced in the last 10 years. By improving data security, privacy, and interoperability, hybrid blockchain technology could significantly renovate the healthcare information exchange industry. Utilizing the advantages of both public and private blockchains [39], a hybrid strategy provides healthcare institutions with an adaptable and expandable solution. Overview of steps to integrate blockchain in healthcare is given in Fig. 3.

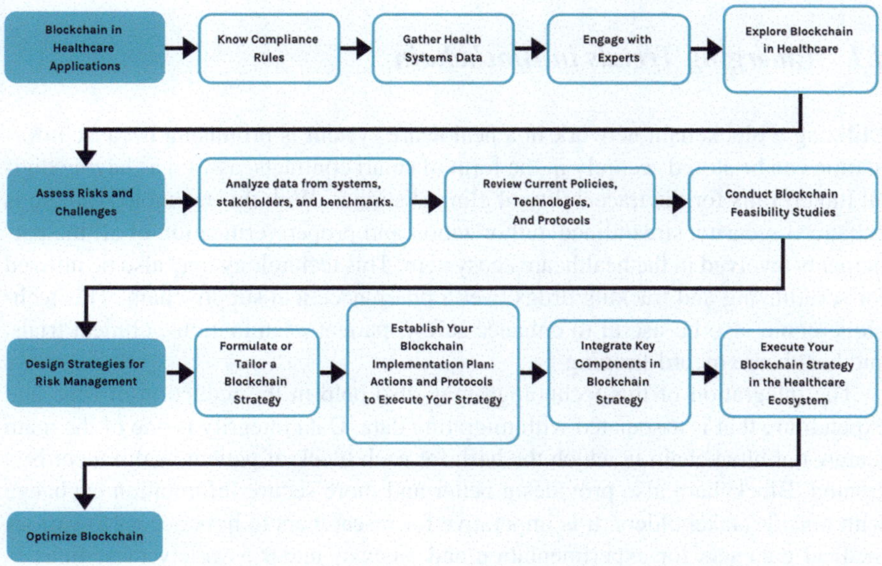

Fig. 3 Integration steps of blockchain in healthcare

4 Case Studies and Real-World Implementations

Blockchain technology has been adopted increasingly in healthcare, with frequent case studies and implementations in the real world identifying its potential. For example, the MediLedger Network [40] influences blockchain to enrich transparency and legitimacy in the medicinal supply chain, reducing the hazard of forgery medications. Likewise, Estonia's eHealth system [41] incorporates blockchain to manage records in healthcare securely, thereby enabling unified access and preserving data privacy. Another prominent application of blockchain in healthcare is Guardtime Health [42], in which employees utilise blockchain to confirm data integrity and streamline procedures for healthcare workers. These case studies highlight how blockchain is renovating healthcare by addressing serious challenges to ensure security, efficacy, and belief. However, benefits of supply chain frequently hamper seamless partnership and integration. Additionally, experts' express confidence about the mounting implementation of digital technologies. Integrating blockchain with corresponding tools like IoT and AI offers major potential to improve traceability, and faith within the supply chain. A digitalized supply chain model leveraging these integrated technologies ensures healthier information availability and modernizes communication across the entire medicinal supply chain. The integration of these technologies will create a more secure, transparent, and efficient pharmaceutical supply chain, ultimately safeguarding public health and trust. In conclusion, while blockchain and related technologies present gifted resolutions for addressing interoperability challenges, their wider implementation joints on overcoming applied implementation barriers.

4.1 Emerging Trends in Blockchain

Utilizing a blockchain network in a healthcare system is promising because information can be stored securely in the form of smart contracts, as well as having digital fingerprints for the traceability of clinical studies. With this technology, there is enhanced security, streamlined authorization and proper verification of all the participants involved in the healthcare ecosystem. This technology may also be utilized for scrutinizing and tracking drugs over a pharmaceutical supply chain. This technique could also be useful to enhance safety, patient examinations, clinical trials, and healthcare record-keeping.

The integration of this technology may also help in the reduction of time and expenditure that is associated with migrating data. Data integrity is one of the main features of blockchain in which the hash for each block of patient health records is created. Blockchain also provides a better and more secure information exchange with outside stakeholders. It is imperative for researchers to have access to various medical data sets for experimentation and analysis under a variety of settings in clinical trials. Hence, blockchain provides security over data manipulation and tampering with evidence by researchers to affect the outcomes. In a scenario where pharmaceutical companies document results to have an edge over other companies, this technology guarantees transparent and fair clinical trials. Blockchain provides a mechanism to have a transparent, safe, and optimal way of documenting these clinical experiments.

This technology is also ideal for maintaining medical records. This could solve various issues related to handling insurance, exchange of medical information, and various administrative responsibilities. EHRs are usually distributed over various healthcare facilities for a particular patient. A blockchain integration will help combine this segregated data and keep the records at one place. This will provide full access to the patient's medical history and assist healthcare professionals in making informed decisions. This technology provides guaranteed privacy and tampering-free records of a patient. It also solves the problem of inconsistent data as well as made-up results in clinical trials. These trials are unbiased and trustworthy, thanks to the integration of blockchain into the healthcare ecosystem. There have been various issues related to the validity of drugs, which was a direct consequence of the ability to trace a drug supply chain. The blockchain provides the transparency and immutable record-keeping of drugs from manufacturing to consumption. Consequently, there is no scope for fake drugs to enter the market. Because of its interoperability, clinicians can easily access comprehensive medical records, which help with diagnosis and makes treatment plans more accurate.

Blockchain encourages transparency and lowers the possibility of content fraud. It becomes simpler to validate clinical research for both clients and participants. For getting authorization and making sure that protocol documentation and results are publicly and transparently verifiable, smart contracts are perfect. The public may now keep a close eye on clinical studies, thanks to this technology. In the end, blockchain is user-centered, giving patients safe, instant access to health and insurance records [43].

4.2 Predictive Analytics for Disease Outbreaks Using Blockchain-Secured Epidemiological Data

From mild sickness to serious impairment and death, infectious illnesses can have a wide range of effects. In addition to upsetting everyday routines and placing a burden on healthcare systems and resources, outbreaks can have significant social and economic effects. Poor hygiene, a lack of access to sanitary facilities and clean water, overcrowding, international travel and commerce, and the emergence of germs resistant to antibiotics are some of the factors that contribute to the spread of illness. A global catastrophe was sparked by the recent COVID-19 epidemic [44], which affected schools, social gathering places, sports facilities, airports, and logistics across several nations. Wide-ranging effects of these disruptions include a worsening of the economic downturn.

Forecasting future occurrences and recognizing several events happening at the same time might motivate people to take preventative action to slow the development of an epidemic. Infectious illness surveillance has been studied recently using statistical, computational, and mathematical methods [44]. These models frequently predict the course of epidemics using time-dependent differential equations. Drawing largely on statistical principles, time-series prediction and observation sequences, including meta-predictors and artificial neural network (ANN)-based approaches, are used using empirical COVID-19 data. An important benefit of ANNs over certain other machine learning methods is their capacity to automatically extract pertinent characteristics from raw data [8]. When considering accuracy, speed, performance, latency, and computing cost, among other performance measures, ANNs provide effective outcomes. For public results and conclusions based on recorded statistical data to be reliable, data authenticity and validation are essential in epidemic surveillance. As a result, the creation of tracking apps has become crucial for preserving data consistency and integrity as well as stopping virus spread. Combating a viral disease requires an approach which involves various facets like hygiene, vaccination, clean water, and proper access to sanitary facilities. There has to be a coordinated and efficient collaboration among various healthcare officials and other significant departments of a community to properly identify and have a response to contain the disease. Blockchain technology could prove promising in providing a transparent and safe contactless tracking system. This would be helpful in collaboration as well as stopping the spread of the disease by identifying various infectious clusters to track the motion of affected people and their immediate contacts.

This technology would also be useful in providing a reliable and immutable way of monitoring the delivery of vaccines. It is imperative to contain the fraud and sell vaccines on the black market. It also ensures proper distribution of vaccines to the needy. The same monitoring mechanism using blockchain could be established for medical equipment and supplies. In short, this technology has tremendous potential for creating a safe distributed network for exchanging information and developing new ideas for therapy and vaccines. Consequently, this shall improve the efficacy and efficiency of efforts to combat contagious diseases and help discover vaccines and strategies promptly [44].

5 Why Use Blockchain?

Blockchains work on incentive systems for sharing genetic data through tokens or cryptocurrency. In this sense, "fairness" refers to the monetary gains that data owners obtain from sharing their information, which enables them to take part in scientific studies or other data applications without giving up all control. Additionally, as shown in [45], incentive systems can compensate network nodes for doing certain tasks, including sequence (HTS) read mapping. As a result, anybody or any organization may join the network, do analysis, and obtain redeemable tokens without restriction.

One important characteristic of blockchain networks that is often emphasized is their decentralized structure. To accomplish this decentralization, consensus methods are essential since they allow network nodes to come to agreements independently of a centralized authority. Depending on the application, decentralization has different advantages. For instance, [46] shows how decentralized open access might be achieved by integrating blockchain technology with the InterPlanetary File System (IPFS). Medical research and development, particularly during illness outbreaks like COVID-19, depends on the timely transfer of medical resources. Coordination of analysis tasks carried out across several nodes or locations is another benefit of decentralization. Blockchain can mediate machine learning processes and connect global models without a central server [47–51]. In addition to avoiding the single point of failure and control that come with utilizing a third party for coordination, this also eliminates the possibility that the third party may jeopardize data privacy by looking at aggregated information. A comparable method is described in [51], where blockchain is utilized to organize GWAS research and guarantee the legitimacy and confirmation of every operation (transaction) within the decentralized network. The patient, who is the data owner, or dependable third parties working on their behalf, including physicians, should ideally have ownership over genetic data. In today's centralized systems, monitoring access and obtaining required consent may be difficult and time-consuming. According to a number of studies, the main reason people use blockchain is for data management, especially in genetic markets where the goal is to give people more control over data access and use. To let patients have complete authority over their data, several proposals [52] include storing patient permission on the blockchain.

Often mentioned in the reviewed literature, immutable and tamper-proof data storage is a widely desired feature of blockchains. Data loss and tampering are prevented by blockchain's immutability, which is essential for the majority of genomic applications. The cryptographic hash points that form the basis of the tamper-proof structure guard against both inadvertent and deliberate data manipulation. Any modifications to verified blocks result in discrepancies that may be identified by any network node, guaranteeing a trustworthy and consistent shared ledger between parties that are either untrusted or semi-trusted. Because immutability applies to on-chain data that is shared across all nodes and may be viewed if unencrypted, data privacy must be carefully taken into account. A number of features are made possible via smart contracts, including as policy enforcement, access control, and token creation and distribution. Many papers include smart contracts as a fundamental part of their proposals, even though not all blockchain systems or applications need them. Commercial platforms for exchanging genomic data use smart contracts to create tokens that encourage people to give their genetic information. Another important justification for utilizing blockchain is the ability to enforce access rules via smart contracts.

These blockchain characteristics were specifically mentioned in a small number of studies as the main justification for its use. Certain applications, like the online model learning described in [48], which necessitates excellent data availability for all network nodes, depend heavily on availability and reliability. Applications that notify data owners about data access and usage are the main ones that prioritize transparency and traceability. Trust may be established and data owners empowered by utilizing these blockchain capabilities. For instance, [52] emphasized the value of openness, contending that when patients are aware of how their data is being utilized, they are more inclined to provide their genetic information for study.

5.1 Fraud Detection in Insurance Claims Using Blockchain-Based Transaction Records

The epitome of patient care and the financial stability of insurance companies are seriously threatened by healthcare fraud. In addition to resulting in monetary losses, these fraudulent acts endanger patient safety by possibly causing injury through faked or needless medical treatments [53]. The majority of current fraud detection techniques rely on rule-based algorithms and manual audits, which frequently fail to identify intricate fraudulent schemes. Data quality, scalability, and the incapacity to adjust to changing fraud strategies are some of the problems that plague traditional systems [54]. These sophisticated methods still face several obstacles, though, such as high computing requirements, data privacy issues, and the difficulty of deciphering intricate models [55, 56]. Blockchain technology provides a fresh approach to these problems. For processing medical insurance claims, its decentralized and unchangeable structure offers a safe, open, and impenetrable platform [57]. The suggested solution uses Ethereum's blockchain platform and smart contracts to:

- Generate a tamper-proof ledger for every claim.
- Validate claims in real-time.
- Use blockchain to verify the validity of claims.
- Detect fraudulent claims in real-time.

In the peer-to-peer structure of the blockchain network, every member keeps a copy of the shared ledger. To ensure data integrity and authenticity, this ledger can only be added to but not altered [57]. Researchers are increasingly using Ethereum and Hyperledger, the two primary blockchain technologies, in healthcare insurance fraud detection. To manage insurance procedures, [58] promoted Ethereum's scalability and security characteristics, highlighting its adherence to healthcare regulations and its ability to perform thousands of transactions per second while protecting user data via its permissioned network. Ethereum is especially well-known for its capacity to solve current system issues, such as utilizing EDI (Electronic Data Interchange) validators to confirm insurance data and adhering to HIPAA (Health Insurance Portability and Accountability Act) regulations [59]. Various researchers have shown how insurance fraud and its detection is one of the advantages of using hyperledger. Consequently, there are built-in capabilities of this trending technology to efficiently identify and help in the reduction of fraudulent claims [56–59].

Even if there are any fraudulent claims, the data is recorded on the blockchain, which can never be rectified. The healthcare professional has no means to modify any false claim. The detection of fraud is enhanced by utilizing smart contracts. It is a way to automate the process of validating claims and flagging any questionable transactions. A blockchain-based mechanism for fraud detection is also a possibility. To detect any false claims, there is a need for pre-determined standards for smart contracts with prerequisite circumstances and rules. It also gives insurance organizations the ability to automatically assess and verify claims by using smart contracts. This technology guarantees compliance promptly and reduces the possibility of fraud [60].

6 Challenges and Solutions for Blockchain Implementation in Healthcare

As far as the issues related to healthcare information exchange are concerned, privacy, security compatibility, and reduction in costs are at the top of the list. There are also the dangers of identity theft, data breaches, and unauthorized access to information when the information is shared on a network. To create more secure HIE systems, recent research has concentrated on cutting-edge technology, like blockchain and artificial intelligence. Blockchain is seen as a ground-breaking technology that offers a distributed and decentralized environment that can overcome the limits of HIE and EHR administration while delivering a more safe, secure, and decentralized setting. An additional layer of security is provided by AI, which is essential for identifying suspicious transactions in blockchain networks. Governments and the academic community must thus give careful consideration to creating smart HIE systems that integrate both technologies. Figure 4 provides the summary of challenges.

Blockchain holds a strong potential for improving healthcare by integrating security, transparency, and patient-centric care. However, its implementation is associated with several challenges given as under:

6.1 Technical Challenges

Given that healthcare companies frequently employ many EHR systems, each with its own data formats, standards, and protocols, interoperability poses a considerable problem. This results in problems like disparities in coding schemes and data formats. For instance, there may be differences in how patient data is interpreted if one hospital utilizes SNOMED CT for diagnosis while another employs ICD-10 coding.

Fig. 4 Challenges of blockchain in healthcare

Electronic health records are one of the important entities in a healthcare system which is transmitted over a network. There must be safeguards for providing secure and fast information exchange. Cybercriminals are always waiting for opportunities to exploit and breach or get illegal access to the EHRs. There must be some robust access control and authorization mechanisms to guarantee the legitimacy of users. Techniques like end-to-end encryption could also be implemented for a safe and secure journey from the source to the destination. There should be mechanisms for audio tracking and cryptography as well. The data shall also be consistent, correct, and complete with no errors or omissions that could result in accurate diagnosis or improper treatments [61].

6.2 *Regulatory and Legal Challenges*

Legal and regulatory concerns are some of the most important issues that must be addressed. There are various scenarios where regulations are needed. For example, patient consent, sharing of data and agreements, data retention and breach notifications, etc. These legal issues give rise to various accountability issues. One of the critical components in a healthcare system is getting the consent of the patient before any of his data can be shared. Currently, there are no concrete regulations or standards for getting such consent. There are also concerns related to the rights of health data in a healthcare ecosystem. This is one of the most important and certain issues which needs a proper solution [61].

6.3 Ethical Considerations

In the recent past, there has been an increase in the number of attacks on private health information [62]. To find a solution to data security and confidentiality, all the stakeholders of a healthcare system must work together to provide solutions to ethical and legal issues [63]. These dimensions have been thoroughly examined in earlier research, which has brought to light several important problems. Ensuring the security of medical data and safeguarding patient privacy are major ethical concerns. Large amounts of patient data, including private and medical information, are frequently used by AI in the healthcare industry. Although strict data protection laws are established by laws like the GDPR in Europe and the HIPAA in the USA, the quick development of AI technologies frequently surpasses these legal frameworks, posing risks. To counter data breaches, many studies highlight the necessity of strong data encryption, anonymization methods, and stringent access restrictions.

The caliber of the training data determines how equitable AI systems are. AI systems have the potential to unfairly treat particular patient populations by reinforcing or even exacerbating pre-existing biases in medical studies. Biases in training data sets can result in illegal discrimination, especially against minority groups, as studies by [64, 65] have shown. Diverse and representative data are needed to address these biases, and algorithms must be continuously monitored and regulated to guarantee fairness and impartiality in the provision of contemporary healthcare. Concerns regarding autonomy and trust are raised by the fact that patients and healthcare professionals frequently find it difficult to comprehend how these systems make decisions. According to studies by [66], explainable AI is essential for the decision-making process to be comprehensible and open to human review. To preserve patient confidence and make sure that they can make knowledgeable healthcare decisions, this openness is essential [67].

6.4 Blockchain into Healthcare: Integration Problems and Their Resolutions

There are several obstacles to overcome when putting blockchain and AI-based frameworks into practice. Several elements, including implementation tactics, are related to practical issues. Decentralized and centralized [68] techniques are distinguished in the literature. In decentralized approaches, patient information is kept locally at each health organization or on-site and exchanged as requested. This method lowers data access latency while giving each business complete control and administration over servers, infrastructure, and security regulations. Decentralized approaches, however, need a large initial outlay, intricate scalability, and increased maintenance accountability. On the other hand, centralized solutions keep patient data in cloud service provider-managed data centers. This method offers substantial processing power for training and testing intricate AI models and is beneficial for adding blockchain nodes for new entities. Because cloud providers take care of software upgrades, hardware maintenance, and other IT administration duties, centralized approaches also provide a more workable way to incorporate blockchain and AI into health systems. However, this method depends on the provider's policies and presents security and privacy issues.

Blockchain and AI implementation in healthcare requires significant resources in a number of domains [69]. Advanced software and hardware, as well as continuous maintenance, demand a large financial commitment. Competent individuals, such as blockchain developers, AI specialists, data scientists, and healthcare IT specialists, are equally important. To enable seamless technology adoption and upskill current employees, training programs are also required. Large healthcare data sets require a strong data infrastructure, which includes high-speed networks and safe storage options. Navigating healthcare data privacy rules requires legal skills to comply with requirements. Specialized implementation expertise may be obtained by collaborating with technology consultants and suppliers. Figure 5 provides comparison of healthcare system with and without Blockchain technology.

To make sure these technologies successfully satisfy clinical demands and regulatory criteria, test projects and ongoing scrutinizing are essential for assessing functioning, resolving problems, and improving systems [70, 71]. The main integration application and issues are discussed as under:

- The ability of an HIE system to support an increasing number of healthcare institutions is referred to as scalability. For instance, the COVID-19 pandemic required a sharp rise in the number of participating health institutions and organizations, as well as a commensurate requirement for the sharing of patient data. Regretfully, the incapacity of certain health entities' systems to successfully integrate and grow caused them to struggle [72]. By including more health entities in the consortium, horizontal scalability (also known as "scaling out") increases the system's ability to manage more transactions and divide the effort among several networks. However, because there are more integrated nodes and all nodes must be in agreement on the state of the blockchain, horizontal scalability may result in higher communication overhead inside the blockchain network [73, 74]. Moreover, the computational complexity of AI models may grow due to the higher transaction volume [75]. Vertical scaling is the process of enhancing each health entity's capability through hardware upgrades (e.g., more powerful CPUs, more memory). Although this might increase network bandwidth, storage capacity, and processing power to manage more data and transactions, these improvements can be costly and might not be practical for all organizations in a decentralized health network [76]. A single point of failure is introduced by vertical scaling; the network as a whole may be severely impacted by the failure of a strong node [77, 78]. Researchers have suggested a number of tactics, such as mini-blockchains, caching systems, tokenization, sharding, the Chord algorithm, and short-term data sharing, to overcome these scaling issues. The authors of [79] give a summary of several scaling options and divide them into two primary groups: blockchain redesign and storage optimization.

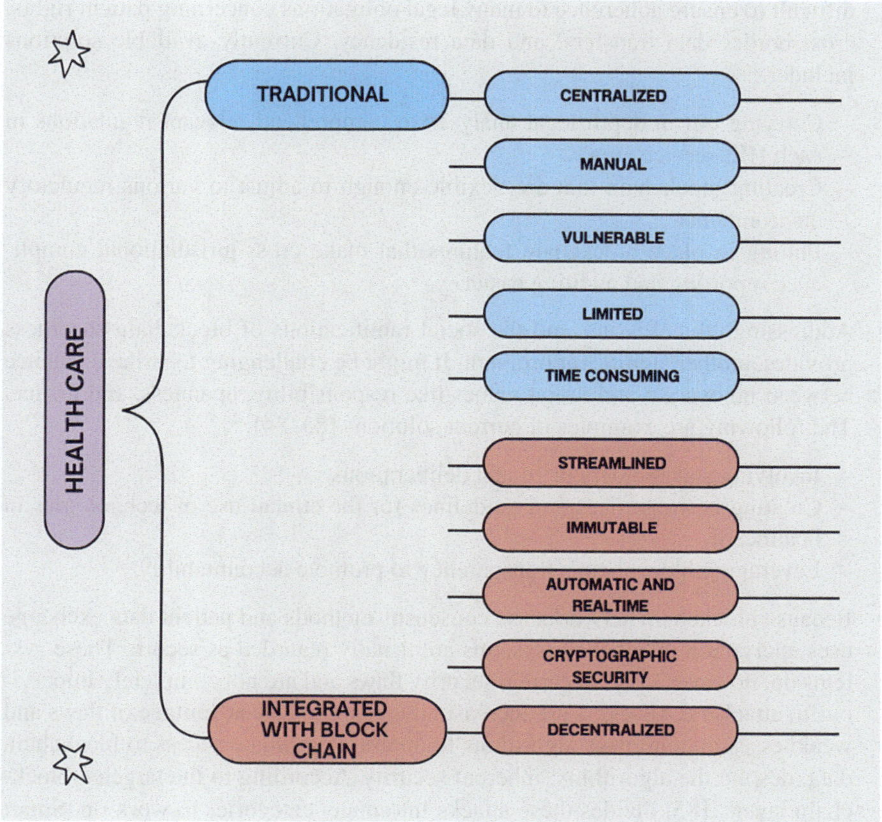

Fig. 5 Healthcare with and without blockchain

- Interoperability, which ISO3 [80] defines as effective communication across many domains without compromising the integrity of dispersed electronic health information, is crucial for the adoption of HIE systems. With the capacity to include novel models and methods for information exchange inside HIE systems, the majority of suggested AI and blockchain solutions prioritize patient-driven interoperability. According to the literature, there are two primary categories of interoperability issues in HIE systems, both of which have been well studied and for which effective fixes have been suggested. The techniques and models used to create interoperable HIE systems are referred to as semantic interoperability. An overview of several standards and solutions for resolving semantic interoperability issues may be found in [81]. A classification of syntactical interoperability frameworks and solutions suggested is provided in [82]. These include technical message creation, knowledge object models and representational forms (UML and XML).
- Navigating the many healthcare legislation and data protection rules across several countries is a major barrier when implementing blockchain systems. It is

difficult to ensure adherence to many legal obligations concerning patient rights, cross-border data transfers, and data residency. Currently available solutions include:

 - Carrying out in-depth legal analyses to comprehend relevant regulations in each HIE service area.
 - Creating blockchain that are flexible enough to adjust to various regulatory environments.
 - Putting in place blockchain features that make cross-jurisdictional compliance reporting and auditing easier.

- Addressing ethical issues and the social ramifications of blockchain sheerness provides another significant problem. It might be challenging to strike a balance between innovation and moral values like responsibility, openness, and justice. The following are examples of current solutions [83, 84]:

 - Involving stakeholders in ethical deliberations.
 - Creating best practices and guidelines for the ethical use of technologies in healthcare.
 - Leveraging blockchain's transparency to promote accountability.

- Because blockchain networks use consensus methods and patient data exchange uses encryption, blockchain systems are usually regarded as secure. These systems do, however, contain certain security flaws and are not completely impervious to attackers. Attackers are increasingly trying to take advantage of flaws and weaknesses in consensus algorithms to obtain illegitimate access to blockchain data, despite the algorithms' inherent security. According to the targeted blockchain layers, [85] divides these attacks into many categories to work on. Smart contracts frequently work with data storage that is both on-chain and off-chain, where data dependability isn't always assured. Consequently, one of the most important tasks in HIE systems is creating smart contracts. Researchers have lately put out several mitigation techniques and methods, including Oyente, Town Crier, ekiden, and Enigma, to address smart contract vulnerabilities [85]. Implementing top guidelines in smart contract creation, conducting frequent audits, to finding and fix vulnerabilities early are the main goals of these tactics.
- Human aspects are important to take into account while deploying blockchain systems in healthcare, much like implementation costs. Because they are accustomed to the current systems or are afraid of losing their jobs, healthcare workers may be reluctant to adopt new techniques. This hesitancy might hinder the overall efficacy of incorporating blockchain technology into HIE systems and cause deployment delays. Health organizations should involve employees early in the process to get their input and solve any issues to address this. Various workshops and seminars may be organized for staff members to learn about blockchain technology. There must be proper training programs given to healthcare professionals to provide technical expertise. The training shall encompass technical as well as practical implementation of blockchain technology to reduce errors, inefficiency, and user dissatisfaction.

7 Conclusion

Blockchain is a decentralized and hyperledger-based technology that provides a foundational mechanism for a secure and efficient healthcare system. With the rising issues of data breaches and privacy violations, blockchain guarantees data integrity and unauthorized access to patient information. It offers improved interoperability for all the stakeholders in a healthcare ecosystem. There are concerns related to data silos which hinder the smooth transmission of patient information and result in inefficiencies and medical mistakes. This technology can provide a safe and shared platform for information exchange and a robust mechanism to complete patient records. As far as the research and clinical trials are concerned, blockchain provides enhanced data quality and transparency. The technology also inhibits any unbiased edge given to a particular company by documenting biased results. Blockchain technology also provides traceability of drugs on a pharmaceutical supply chain which not only makes fake drugs impossible to enter the market but also makes stakeholders accountable. It could help in providing patient-centered and data-driven healthcare while catering to the issues of management, security, and data privacy. The integration is still not ripe but continuous efforts would be beneficial for better and efficient healthcare.

References

1. National Academics of Sciences, Engineering, and Medicine. (2018). *Crossing the global quality chasm: Improving healthcare worldwide*. The National Academics Press.
2. Powell, M. (2023). Learning lessons from the COVID-19 pandemic. *Health Economics, Policy and Law, 18*(1), 88–103.
3. da Silva, S. J. R., do Nascimento, J. C. F., Germano Mendes, R. P., Guarines, K. M., Targino Alves da Silva, C., da Silva, P. G., et al. (2022). Two years into the COVID-19 pandemic: Lessons learned. *ACS Infectious Diseases, 8*(9), 1758–1814.
4. Aitken, J., Esain, A. E., & Williams, S. (2021). Management of complexity in the care ecosystem. *Supply Chain Management: An International Journal, 26*(4), 481–494.
5. Fries, J. F., Koop, C. E., Sokolov, J., Beadle, C. E., & Wright, D. (1998). Beyond health promotion: Reducing need and demand for medical care: Health care reforms to improve health while reducing costs. *Health Affairs, 17*(2), 70–84.
6. Viswanadham, N. (2021). Ecosystem model for healthcare platform. *Sādhanā, 46*(4), 188.
7. Jaleel, A., Mahmood, T., Hassan, M. A., Bano, G., & Khurshid, S. K. (2020). Towards medical data interoperability through collaboration of healthcare devices. *IEEE Access, 8*, 132302–132319.
8. Ravaghi, H., Afshari, M., Isfahani, P., & Bélorgeot, V. D. (2019). A systematic review on hospital inefficiency in the Eastern Mediterranean Region: Sources and solutions. *BMC Health Services Research, 19*, 1–20.
9. Smith, E., & Eloff, J. H. (1999). Security in health-care information systems—Current trends. *International Journal of Medical Informatics, 54*(1), 39–54.
10. World Health Organization. (2009). *Improving hospital performance in the Eastern Mediterranean Region*. WHO.

11. Jennath, H. S., Anoop, V. S., & Asharaf, S. (2020). *Blockchain for healthcare: Securing patient data and enabling trusted artificial intelligence* (Vol. 6, pp. 15–23).
12. Koutkias, V. (2019). From data silos to standardized, linked, and FAIR data for pharmacovigilance: Current advances and challenges with observational healthcare data. *Drug Safety, 42*(5), 583–586.
13. Rather, I. H., & Idrees, S. M. (2021). Blockchain technology and its applications in the healthcare sector. In *Blockchain for healthcare systems* (pp. 17–25). CRC Press.
14. Arza, M. S., & Panda, S. K. (2022). An integration of blockchain and machine learning into the health care system. In *Machine learning adoption in blockchain-based intelligent manufacturing* (pp. 33–58). CRC Press.
15. Cheng, A. S., Guan, Q., Su, Y., Zhou, P., & Zeng, Y. (2021). Integration of machine learning and blockchain technology in the healthcare field: A literature review and implications for cancer care. *Asia-Pacific Journal of Oncology Nursing, 8*(6), 720–724.
16. Rajeswari, B., Pujitha, P., Kumar, G. P., & Syamala, M. K. S. (2024). Health care and management using block chain and machine learning. *Health Care, 13*(2), 241–246.
17. Idrees, S. M., & Nowostawski, M. (Eds.). (2024). *Blockchain transformations: Navigating the decentralized protocols era*. Springer.
18. Mettler, M. (2016). Blockchain technology in healthcare: The revolution starts here. In *Proc. 2016 IEEE 18th International Conference on e-Health Networking Applications and Services (Healthcom), Sept. 14–16* (pp. 1–3). IEEE.
19. Saindane, P., Jethani, Y., Mahtani, P., Rohra, C., & Lund, P. (2020, October). Blockchain: A solution for improved traceability with reduced counterfeits in supply chain of drugs. In *2020 International Conference on Electrotechnical Complexes and Systems (ICOECS)* (pp. 1–5). IEEE.
20. Trautmann, L., Hübner, T., & Lasch, R. (2024). Blockchain concept to combat drug counterfeiting by increasing supply chain visibility. *International Journal of Logistics Research and Applications, 27*(6), 959–985.
21. Musamih, A., Salah, K., Jayaraman, R., Arshad, J., Debe, M., Al-Hammadi, Y., & Ellahham, S. (2021). A blockchain-based approach for drug traceability in healthcare supply chain. *IEEE Access, 9*, 9728–9743.
22. Pinto, R. P., Silva, B. M., & Inacio, P. R. (2022). A system for the promotion of traceability and ownership of health data using blockchain. *IEEE Access, 10*, 92760–92773.
23. RBC Capital Markets. Navigating the changing face of healthcare episode. Retrieved October 30, 2023, from https://www.rbccm.com/en/gib/healthcare/story.page
24. Lee, H.-A., Kung, H.-H., Udayasankaran, J. G., Kijsanayotin, B., Marcelo, A. B., Chao, L. R., & Hsu, C.-Y. (2020). An architecture and management platform for blockchain-based personal health record exchange: Development and usability study. *Journal of Medical Internet Research, 22*, e16748.
25. Carlos Ferreira, J., Elvas, L. B., Correia, R., & Mascarenhas, M. (2024). Enhancing EHR interoperability and security through distributed ledger technology: A review. *Healthcare (Basel), 12*(19), 1967.
26. Selvamuthu, C. M., Lavaraju, B., & Sundaram, A. (2024, October). A novel approach of streamlining claims processing and fraud prevention in health insurance through blockchain technology. In *2024 8th International Conference on I-SMAC (IoT in Social, mobile, Analytics and Cloud)* (pp. 611–618). IEEE.
27. Kadhum Idrees, A., Alhussein, D. A., & Harb, H. (2023). Energy-efficient multisensor adaptive sampling and aggregation for patient monitoring in edge computing based IoHT networks. *Journal of Ambient Intelligence and Smart Environments, 15*, 235–253.
28. Idrees, S., & Nowostawski, M. (2022). *Transformations through blockchain technology*. Springer.
29. Merhej, J., Harb, H., Abouaissa, A., & Idoumghar, L. (2023). DeepChain: A deep learning and blockchain based framework for detecting risky transactions on HIE system. In *Proceedings*

of the 2023 IEEE International Conference on enabling technologies: Infrastructure for collaborative enterprises (WETICE), Paris, France, 14–16 December 2023 (pp. 1–6). IEEE.
30. Ali, S., Abdullah, Armand, T. P. T., Athar, A., Hussain, A., Ali, M., et al. (2023). Metaverse in healthcare integrated with explainable AI and blockchain: Enabling immersiveness, ensuring trust, and providing patient data security. *Sensors, 23*, 565.
31. Mukherjee, P., Pradhan, C., Tripathy, H. K., & Gaber, T. (2023). KryptosChain—A blockchain-inspired, AI-combined, DNA-encrypted secure information exchange scheme. *Electronics, 12*, 493.
32. Borna, S., Maniaci, M. J., Haider, C. R., Maita, K. C., Torres-Guzman, R. A., Avila, F. R., et al. (2023). Artificial intelligence models in health information exchange: A systematic review of clinical implications. *Healthcare, 11*, 2584.
33. Mozumder, M. A. I., Armand, T. P. T., Imtiyaj Uddin, S. M., Athar, A., Sumon, R. I., Hussain, A., & Kim, H. C. (2023). Metaverse for digital anti-aging healthcare: An overview of potential use cases based on artificial intelligence, blockchain, IoT technologies, its challenges, and future directions. *Applied Sciences, 13*, 5127.
34. Jadav, D., Jadav, N. K., Gupta, R., Tanwar, S., Alfarraj, O., Tolba, A., et al. (2023). A trustworthy healthcare management framework using amalgamation of AI and blockchain network. *Mathematics, 11*, 637.
35. Sittig, D. F., Wright, A., Coiera, E., Magrabi, F., Ratwani, R., Bates, D. W., & Singh, H. (2020). Current challenges in health information technology-related patient safety. *Health Informatics Journal, 26*, 181–189.
36. Murugan, A., Chechare, T., Muruganantham, B., & Kumar, S. G. (2020). Healthcare information exchange using blockchain technology. *International Journal of Electrical and Computer Engineering, 10*, 421.
37. Awotunde, J. B., Bhoi, A. K., & Barsocchi, P. (2021). Hybrid cloud/fog environment for healthcare: An exploratory study, opportunities, challenges, and future prospects. In *Hybrid artificial intelligence and IoT in healthcare* (pp. 1–20). Springer.
38. Zhuang, Y., Sheets, L., Shae, Z., Tsai, J. J., & Shyu, C. R. (2018). Applying blockchain technology for health information exchange and persistent monitoring for clinical trials. In *Proceedings of the AMIA Annual Symposium Proceedings, San Francisco, CA, USA, 3–7 November 2018* (p. 1167). American Medical Informatics Association.
39. Ahmed Teli, T., & Masoodi, F. (2021, July 8). Blockchain in healthcare: Challenges and opportunities. In *Proceedings of the International Conference on IoT Based Control Networks & Intelligent Systems - ICICNIS 2021*. Retrieved from https://ssrn.com/abstract=3882744 or https://doi.org/10.2139/ssrn.3882744
40. Agrawal, D., Minocha, S., Namasudra, S., & Gandomi, A. H. (2022). A robust drug recall supply chain management system using hyperledger blockchain ecosystem. *Computers in Biology and Medicine, 140*, 105100.
41. Kruus, P., Aaviksoo, A., Hallik, R., Uus, M., Maghiros, I., Villalba, E., & Zamora, B. (2013). *Country study Estonia*. European Union.
42. Heston, T. F. (2017). A case study in blockchain healthcare innovation. *International Journal of Current Research, 9*(11), 60587–60588.
43. Haleem, A., Javaid, M., Singh, R. P., Suman, R., & Rab, S. (2021). Blockchain technology applications in healthcare: An overview. *International Journal of Intelligent Networks, 2*, 130–139. https://doi.org/10.1016/j.ijin.2021.09.005
44. Idrees, S. M., Nowostawski, M., & Jameel, R. (2021). Blockchain-based digital contact tracing apps for COVID-19 pandemic management: Issues, challenges, solutions, and future directions. *JMIR Medical Informatics, 9*(2), e25245.
45. Ileri, A. M., Ozercan, H. I., Gundogdu, A., Senol, A. K., Ozkaya, M. Y., & Alkan, C. (2016). Coinami: A cryptocurrency with DNA sequence alignment as proof-of-work. *Computational Engineering, Finance, and Science*. https://doi.org/10.48550/arXiv.1602.03031
46. Glicksberg, B. S., Burns, S., Currie, R., Griffin, A., Wang, Z. J., Haussler, D., et al. (2020). Blockchain-authenticated sharing of genomic and clinical outcomes data of patients with can-

cer: A prospective cohort study. *Journal of Medical Internet Research, 22*(3), e16810. https://doi.org/10.2196/16810
47. Kuo, T.-T. (2020). The anatomy of a distributed predictive modeling framework: Online learning, blockchain network, and consensus algorithm. *JAMIA Open, 3*(2), 201–208.
48. Kuo, T.-T., Gabriel, R. A., Cidambi, K. R., & Ohno-Machado, L. (2020). EXpectation Propagation LOgistic Regression on permissioned blockCHAIN (ExplorerChain): Decentralized online healthcare/genomics predictive model learning. *Journal of the American Medical Informatics Association, 27*(5), 747–756.
49. Kuo, T.-T., Gabriel, R. A., & Ohno-Machado, L. (2019). Fair compute loads enabled by blockchain: Sharing models by alternating client and server roles. *Journal of the American Medical Informatics Association, 26*(5), 392–403.
50. Kuo, T.-T., Kim, J., & Gabriel, R. A. (2020). Privacy-preserving model learning on a blockchain network-of-networks. *Journal of the American Medical Informatics Association, 27*(3), 343–354.
51. Zhang, Y., Zhao, X., Xue, L., Zhong, M., Curtis, C., & Chen, C. (2019). Enabling privacy-preserving sharing of genomic data for GWASs in decentralized networks. In *Proceedings of the 12th ACM International Conference on web search and data mining* (pp. 204–212). ACM.
52. Mamo, N., Martin, G. M., Desira, M., Ellul, B., & Ebejer, J.-P. (2020). Dwarna: A blockchain solution for dynamic consent in biobanking. *European Journal of Human Genetics, 28*(5), 609–626.
53. Ismail, L., & Zeadally, S. (2019). Healthcare insurance frauds: Taxonomy and blockchain-based detection framework (block-HI). *IT Professional, 23*(4), 36–43. https://doi.org/10.1109/MITP.2021.3071534
54. Amponsah, A. A., Adekoya, A. F., & Weyori, B. A. (2022). A novel fraud detection and prevention method for healthcare claim processing using machine learning and blockchain technology. *Decision Analytics Journal, 4*, 100122. https://doi.org/10.1016/j.dajour.2022.100122
55. Kapadiya, K., Patel, U., Gupta, R., Alshehri, M. D., Tanwar, S., Sharma, G., & Bokoro, P. N. (2022). Block-chain and AI-empowered healthcare insurance fraud detection: An analysis, architecture, and future prospects. *IEEE Access, 10*, 79606–79627. https://doi.org/10.1109/ACCESS.2022.3194569
56. Kaafarani, R., Ismail, L., & Zahwe, O. (2023). An adaptive decision-making approach for better selection of block-chain platform for health insurance frauds detection with smart contracts: Development and performance evaluation. *Procedia Computer Science, 220*, 470–477. https://doi.org/10.1016/j.procs.2023.03.060
57. Ncube, N., Mutunhu, B., & Sibanda, K. (2022). Land Registry using a distributed ledger. In *IST-Africa Conference (IST-Africa), Ireland* (pp. 1–7) Retrieved from https://ieeexplore.ieee.org/document/9845584
58. Kapri, P., & Venkatesh, B. (2023). Usage of block-chain technology for tamper-proof audit and managing insurance process. https://doi.org/10.13140/RG.2.2.12389.17122
59. Saldamli, G. (2020). *2020 Seventh International Conference on Software Defined Systems (SDS): Paris, France. April 20–23, 2020.*
60. Mackey, T. K., Miyachi, K., Fung, D., Qian, S., & Short, J. (2020). Combating health care fraud and abuse: Conceptualization and prototyping study of a block-chain antifraud framework. *Journal of Medical Internet Research, 22*(9), e18623. https://doi.org/10.2196/18623
61. Merhej, J., Harb, H., Abouaissa, A., & Idoumghar, L. (2024). Toward a new era of smart and secure healthcare information exchange systems: Combining blockchain and artificial intelligence. *Applied Sciences, 14*, 8808. https://doi.org/10.3390/app14198808
62. Panayides, A. S., Amini, A., Filipovic, N. D., Sharma, A., Tsaftaris, S. A., Young, A., et al. (2020). AI in medical imaging informatics: Current challenges and future directions. *IEEE Journal of Biomedical and Health Informatics, 24*(7), 1837–1857.
63. Christou, C. D., & Tsoulfas, G. (2021). Challenges and opportunities in the application of artificial intelligence in gastroenterology and hepatology. *World Journal of Gastroenterology, 27*, 6191. https://doi.org/10.3748/wjg.v27.i37.6191

64. Obermeyer, Z., & Emanuel, E. J. (2016). Predicting the future — Big data, machine learning, and clinical medicine. *New England Journal of Medicine, 375*(13), 1216–1219. https://doi.org/10.1056/nejmp1606181
65. Kansal, R., Bawa, A., Bansal, A., Trehan, S., Goyal, K., Goyal, N., & Malhotra, K. (2022). Differences in knowledge and perspectives on the usage of artificial intelligence among doctors and medical students of a developing country: A cross-sectional study. *Cureus, 14*, e21434. https://doi.org/10.7759/cureus.21434
66. Mittelstadt, B. D., & Floridi, L. (2016). *The ethics of biomedical big data*. Springer.
67. Kashyap, S., Morse, K., Patel, B., & Shah, N. H. (2021). A survey of extant organizational and computational setups for deploying predictive models in health systems. *Journal of the American Medical Informatics Association, 28*, 2445. https://doi.org/10.1093/jamia/ocab154
68. Haddad, A., Habaebi, M. H., Islam, M. R., Hasbullah, N. F., & Zabidi, S. A. (2022). Systematic review on ai-blockchain based e-healthcare records management systems. *IEEE Access, 10*, 94583–94615.
69. Jabarulla, M. Y., & Lee, H. N. (2021). A blockchain and artificial intelligence-based, patient-centric healthcare system for combating the COVID-19 pandemic: Opportunities and applications. *Healthcare, 9*, 1019.
70. Elangovan, D., Long, C. S., Bakrin, F. S., Tan, C. S., Goh, K. W., Yeoh, S. F., et al. (2022). The use of blockchain technology in the health care sector: Systematic review. *JMIR Medical Informatics, 10*, e17278.
71. Durneva, P., Cousins, K., & Chen, M. (2020). The current state of research, challenges, and future research directions of blockchain technology in patient care: Systematic review. *Journal of Medical Internet Research, 22*, e18619.
72. Baalbaki, H., Harb, H., Jaber, A., Zaki, C., Abou Jaoude, C., Tout, K., & Tannoury, L. (2022). Fighting against COVID-19: Who failed and who succeeded? *Journal of Computer and Communications, 10*, 32–50.
73. Clim, A., Zota, R. D., & Constantinescu, R. (2019). Data exchanges based on blockchain in m-Health applications. *Procedia Computer Science, 160*, 281–288.
74. Antwi, M., Adnane, A., Ahmad, F., Hussain, R., Rehman, M. H. U., & Kerrache, C. A. (2021). The case of HyperLedger Fabric as a blockchain solution for healthcare applications. *Blockchain: Research and Applications, 2*, 100012.
75. Rouhani, S., & Deters, R. (2021). Data trust framework using blockchain technology and adaptive transaction validation. *IEEE Access, 9*, 90379–90391.
76. Saini, A., Zhu, Q., Singh, N., Xiang, Y., Gao, L., & Zhang, Y. (2020). A smart-contract-based access control framework for cloud smart healthcare system. *IEEE Internet of Things Journal, 8*, 5914–5925.
77. Tao, X., & Velasquez, J. D. (2022). Multi-source information fusion for smart health with artificial intelligence. *Information Fusion, 83*, 93–95.
78. Habib, C., Makhoul, A., Darazi, R., & Salim, C. (2016). Self-adaptive data collection and fusion for health monitoring based on body sensor networks. *IEEE Transactions on Industrial Informatics, 12*, 2342–2352.
79. Mazlan, A. A., Daud, S. M., Sam, S. M., Abas, H., Rasid, S. Z. A., & Yusof, M. F. (2020). Scalability challenges in healthcare blockchain system—A systematic review. *IEEE Access, 8*, 23663–23673.
80. Begoyan, A. (2007). *An overview of interoperability standards for electronic health records*. Society for Design and Process Science.
81. De Mello, B. H., Rigo, S. J., da Costa, C. A., da Rosa Righi, R., Donida, B., Bez, M. R., & Schunke, L. C. (2022). Semantic interoperability in health records standards: A systematic literature review. *Health Technology, 12*, 255–272.
82. Umberfield, E. E., Staes, C. J., Morgan, T. P., Grout, R. W., Mamlin, B. W., & Dixon, B. E. (2023). Syntactic interoperability and the role of syntactic standards in health information exchange. In *Health information exchange* (pp. 217–236). Elsevier.

83. Patel, V. (2019). A framework for secure and decentralized sharing of medical imaging data via blockchain consensus. *Health Informatics Journal, 25*, 1398–1411.
84. Wu, C., Xu, H., Bai, D., Chen, X., Gao, J., & Jiang, X. (2023). Public perceptions on the application of artificial intelligence in healthcare: A qualitative meta-synthesis. *BMJ Open, 13*, e066322.
85. Andrew, J., Isravel, D. P., Sagayam, K. M., Bhushan, B., Sei, Y., & Eunice, J. (2023). Blockchain for healthcare systems: Architecture, security challenges, trends and future directions. *Journal of Network and Computer Applications, 215*, 103633.

Reinventing Agriculture: AI and Blockchain in Farming Practices

Maher Alndiwee, Benita Christopher, Rupinder Pal Kaur, Khyati Marwah, Nasvi Kareem, and Roshan Jameel

1 Introduction

The agricultural industry is experiencing a revolution due to the combined power of blockchain technology and artificial intelligence (AI), which has transformed conventional farming techniques into intelligent farming systems. By 2050, it is expected that there will be nearly ten billion individuals on the planet [1], making food security a crucial issue. Leveraging cutting-edge technologies, smart farming offers innovative solutions to these problems, allowing for increased resilience, sustainability, and efficiency in farming methods. AI's predictive modeling and data analysis abilities provide farmers with the resources they need to make intelligent choices, maximize resource use, and boost production. In the same way, blockchain technology has its applications in several industrial domains [2]. It tackles serious issues like food fraud and safety by opening the door for more transparency, traceability, and trust within agricultural supply chains.

This chapter explores how blockchain technology and artificial intelligence may work together to transform agricultural practices. It starts with investigating the various ways artificial intelligence (AI) is being used in agriculture, such as automated technology to streamline farm operations, precision farming to improve resource efficiency, and predictive analytics to increase crop yields. The discussion then turns to blockchain's revolutionary potential to enhance food safety, combat fraud, and assure transparency in agricultural supply chains. To put these developments into perspective, the chapter provides case studies from real-world

M. Alndiwee (✉) · B. Christopher · R. P. Kaur · N. Kareem · R. Jameel
Westford University College, Sharjah, UAE
e-mail: maher.a@westford.org.uk

K. Marwah
DeMont Institute of Management and Technology, Dubai, UAE

S. M. Idrees et al. (eds.), *Blockchain and Machine Learning Innovations*, Signals and Communication Technology,
https://doi.org/10.1007/978-3-031-88095-7_11

applications, such IBM Food Trust and AgriDigital, which demonstrate how blockchain and AI are used in smart farming. Furthermore, a novel framework for vertical agricultural operations optimization "Agentic AI Workflow and Digital Twin" is proposed. The chapter ends with a summary of the most significant achievements and an outlook on how these technologies will contribute to sustainable agriculture in the future.

2 AI in Agriculture: Enhancing Productivity and Sustainability

Over the last decade, AI has become an invaluable tool for enhancing agricultural practices. With the global demand for food rising and natural resources becoming scarcer, AI offers innovative solutions to address these challenges, promoting both efficiency and sustainability. This chapter explores various AI-powered advancements in agriculture—such as predictive analytics, precision farming, and automated machinery—and examines how they are transforming the sector. From optimizing crop yields to reducing resource waste, AI is reshaping modern farming and supporting the growing need for sustainable, resilient practices. In this part of the chapter, we will cover the different aspects of AI used in agriculture in detail.

2.1 AI for Predictive Analytics to Boost Crop Yields

Predictive analytics is certainly one of AI's most widespread programs in agriculture. Through huge-scale information evaluation, predictive fashions provide farmers with selection assist, supporting them make informed selections approximately planting, irrigation, fertilization, and harvesting, ultimately decreasing risks and enhancing productiveness. These fashions examine information from multiple assets, which include climate forecasts, soil health, historic crop yields, and market demand, allowing farmers to optimize their crop yields with minimum risk of loss [3].

For example, AI-pushed tools help farmers pick out the surest planting times via analyzing historic weather styles and modern soil situations. By supplying real-time insights, AI enables farmers to adapt their schedules, which could reduce the hazard of crop failure because of sudden weather adjustments. Additionally, AI models can forecast crop illnesses and pest infestations, allowing farmers to take preemptive measures. This predictive potential is mainly precious in areas wherein climate variability or pest outbreaks threaten agricultural production, making AI an important tool in global efforts to enhance meals safety [4].

As illustrated in Fig. 1, the activities concerned in agriculture have been widely categorized into three regions: pre-harvesting, harvesting, and publish harvesting.

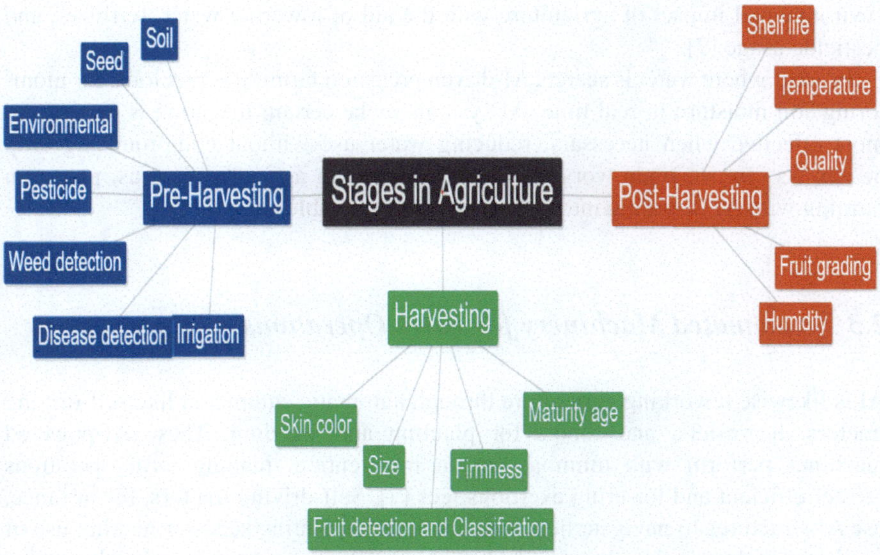

Fig. 1 Important parameters considered in each stage of farming [5]

The improvement of machine learning concept has made extremely good advantages in agriculture. The modern-day technology that's reaping rewards farmers to reduce the losses in the farming with the aid of supplying wealthy tips and insights about the crops [5].

2.2 *Precision Farming for Enhancing Resources Efficiency with AI*

Precision farming is some other vicinity wherein AI significantly influences cutting-edge agriculture. This approach makes use of AI to screen actual-time discipline situations and supply centered interventions, ensuring that inputs like water, fertilizer, and pesticides are carried out precisely and successfully. With AI-powered sensors and drones, farmers can accumulate data on soil moisture, nutrient degrees, and crop health, allowing them to tailor resource use to the unique wishes of different regions in the subject [6]. For instance, sensors equipped with devices gaining knowledge of algorithms discover subject regions desiring extra water or fertilizer, supporting to minimize waste and avoid overuse. AI-equipped drones can also carry out aerial surveys to reveal crop booms, identifying signs and symptoms of stress, sickness, or pest infestations. These drones relay actual-time information to farmers, who can then observe focused remedies rather than blanket programs of chemical compounds. This precise method improves crop yields while lowering the environmental impact of agriculture with the aid of lowering water, fertilizer, and pesticide usage [7].

In areas where water is scarce, AI-driven precision farming is precious. By monitoring soil moisture in real time, AI systems make certain irrigation is carried out most effective when necessary, reducing water use without compromising crop health. As climate trade worsens water shortages in agricultural areas, precision farming with AI is turning into important for sustainable agriculture.

2.3 *Automated Machinery for Farm Operations*

AI is likewise reworking agriculture through automatic equipment like self-driving tractors, harvesters, and robots for planting and weeding. These AI-powered machines perform with minimal human intervention, making farm operations greater efficient and lowering exertions fees [7]. Self-driving tractors, for instance, use AI structures to navigate fields autonomously, planting seeds or making use of fertilizers with pinpoint accuracy. Robotic harvesters can discover and pick out ripe produce, considerably decreasing the need for guide hard work throughout harvest seasons. These machines can be painted around the clock, offering a big productiveness enhance, especially at some point of exertions shortages. Additionally, AI-driven robots can perform repetitive tasks like weeding or thinning plants, which are hard work-intensive and time-ingesting for human people. AI applications including laptop vision and time-series analysis permit farmers to expect most efficient instances for seeding, harvesting, and advertising their crops [8, 9]. AI-primarily based answers contribute to farm profitability and decrease poor environmental affects [10–13]. Research by Koliba and Spett provides examples of machine learning applications in agriculture, as shown in Fig. 2 [14].

The integration of automated machinery into farming not only improves operational efficiency but also enhances the accuracy of farming practices. AI-powered robots can apply pesticides or fertilizers with high precision, ensuring that only the necessary amounts are used. This reduces the environmental footprint of agricultural operations, as fewer chemicals are released into the soil and water systems.

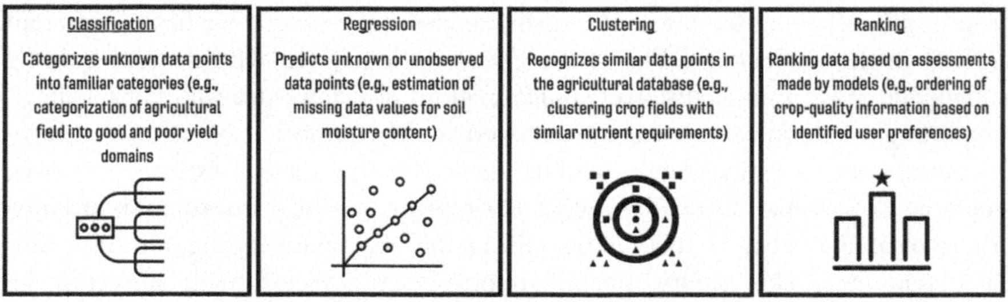

Fig. 2 Machine learning techniques used in agriculture [adapted from Koliba & Spett [14]]

The use of AI in automated machinery is therefore instrumental in promoting both productivity and sustainability in modern agriculture.

3 Blockchain Technology in Agriculture: Enhancing Transparency and Trust

As agriculture will become increasingly integrated into worldwide markets, making sure transparency, traceability, and performance in agricultural supply chains is essential. Blockchain generation has emerged as a pivotal tool in selling transparency throughout the food supply chain, even though numerous barriers—technical, policy-related, regulatory, and educational—still restrict its good-sized adoption in agri-food systems [3]. Blockchain's decentralized and immutable shape gives massive potential to cope with those troubles. Despite being inside the early tiers of adoption within meals supply chains, setting up a framework to evaluate blockchain's influences is critical [15]. This segment discusses how blockchain is transforming agricultural supply chains by using improving food protection, streamlining certification methods, and securing economic transactions. It highlights blockchain's role in fostering agreement by way of recording every step in the food manufacturing and distribution method, ensuring product authenticity, and fighting fraud.

3.1 Blockchain and Transparency in Agricultural Supply Chains

The farm-to-table journey in agriculture involves several intermediaries together with processors, vendors, and shops. Blockchain can reduce the number of middlemen, decreasing transaction charges and boosting earnings margins for farmers. It additionally allows save you fraud, enhances meals protection and great management, and guarantees truthful pricing for farmers. Blockchain's decentralized ledger securely tracks and verifies every step of the supply chain [16]. Key attributes along with traceability, immutability, auditability, and provenance are instrumental in selling supply chain transparency [17]. By imparting actual-time visibility into each stage of production and distribution, blockchain enables stakeholders to verify the authenticity, beginning, and protection of agricultural products [18].

For example, blockchain has been successfully carried out in tracking organic meals production. Consumers frequently pay a top rate for licensed natural merchandise, however verifying those claims can be hard. Blockchain faces every transaction, from planting and harvesting to packaging and transport, on a transparent and immutable ledger, verifying organic certification at every degree [19]. This transparent document addresses customer worries about food safety and product integrity, at the same time as promoting transparency at some stage in the supply chain. Additionally, smart contracts facilitate transactions in the agri-supply chain [20].

3.2 Blockchain and Food Safety

Food safety remains a pressing concern in worldwide agriculture, with foodborne sicknesses posing big dangers to public fitness and causing economic losses. Blockchain can mitigate those risks by way of presenting real-time traceability of food merchandise [21]. In the event of infection or a pandemic, blockchain enables speedy identity of the supply, minimizing the unfolding of harmful merchandise and decreasing the time required for recollects [22].

Smart contracts automatically enforce agreements when predefined conditions are met. For instance, blockchain can automate crop insurance payouts based on adverse weather conditions. Platforms like AgriDigital already use blockchain to manage agricultural commodity sales, ensuring that farmers receive payment immediately upon product delivery, without intermediaries [23]. This helps reduce payment delays and discrepancies, protecting smallholder farmers. As can be seen in Fig. 3, all the entities in the whole supply chain are interconnected in a decentralized blockchain with a smart contract. And every transaction in the whole system can be traced through blockchain network.

3.3 Combatting Fraud and Ensuring Product Authenticity

Fraud is a persistent issue in high-value agricultural sectors, such as organic foods or specialty crops. Blockchain offers a robust solution by creating a tamper-proof digital record, ensuring transparency and verification at each step of the supply chain.

For example, the olive oil industry, which has been prone to adulteration and mislabeling, has begun adopting blockchain to guarantee product authenticity. Blockchain records every step of the production and distribution process, allowing stakeholders to verify product integrity from the olive groves to store shelves [25]. This helps prevent counterfeit products from entering the market, protecting both consumers and producers from fraud.

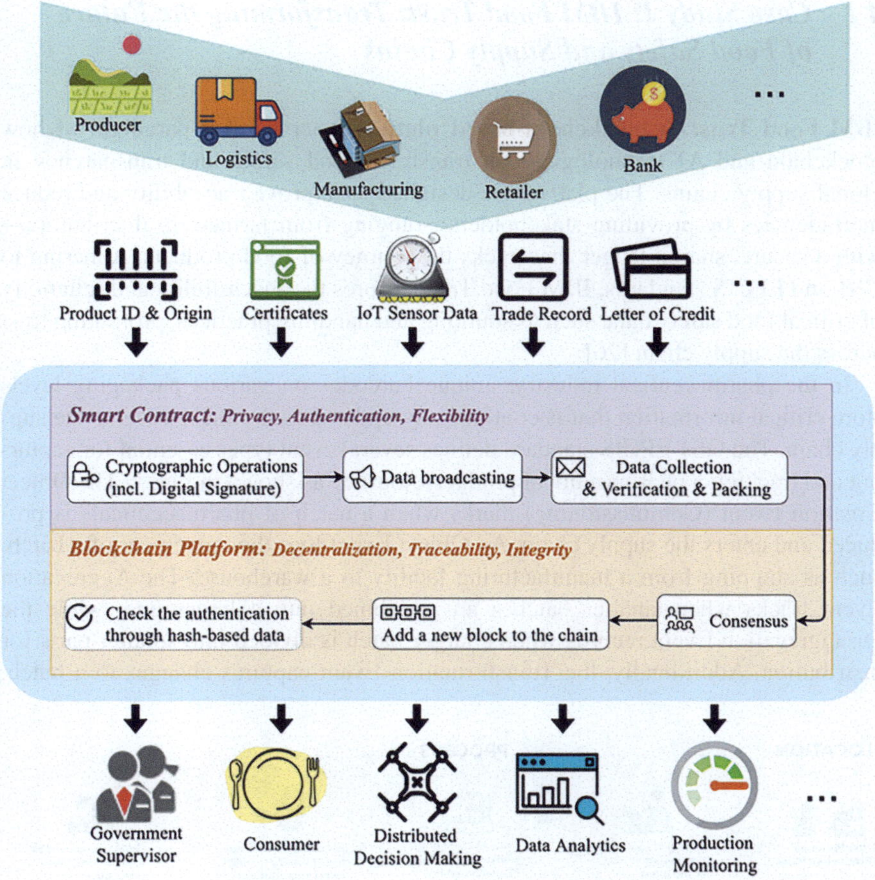

Fig. 3 Agricultural stakeholders, including producers, logistic companies, manufacturers, retailers, and financial institutions, can be involved to create blockchain applications to build up trust of data users, e.g., government supervisors, consumers, and other AI and machine learning-based systems [24]

4 Case Studies: Integrating AI and Blockchain in Modern Agriculture

To illustrate the real-world application of AI and blockchain technologies in agriculture, this section features case studies from diverse farming practices worldwide. These examples demonstrate how the integration of these innovations addresses pressing challenges such as resource management, environmental sustainability, and food security. The following case studies provide practical insights and guidelines for farmers, technologists, and policymakers on how to effectively implement AI and blockchain in agriculture.

4.1 Case Study 1: IBM Food Trust: Transforming the Future of Food Safety and Supply Chains

IBM Food Trust, a blockchain-based platform, serves as a paradigm of how blockchain and AI technologies can transform food safety and transparency in global supply chains. The platform is designed to improve traceability and reduce inefficiencies by providing stakeholders—ranging from farmers to distributors—with a secure, shared ledger that tracks the journey of food products. Adhering to GS1 and EPCIS standards, IBM Food Trust ensures the accessibility and reliability of critical food safety data, such as sourcing and handling practices, promoting trust across the supply chain [26].

In the pharmaceutical industry, unique barcodes on various packaging levels store critical information that is continuously updated as products traverse the supply chain. The GS1 EPCIS standard defines several event types essential for capturing and tracking a pharmaceutical product's journey as shown in Fig. 4. The Object Creation Event (Commissioning) marks when a batch of pharmaceuticals is produced and enters the supply chain. An Object Event logs the movement of a batch, such as shipping from a manufacturing facility to a warehouse. The Aggregation Event tracks when smaller batches are combined into a larger one, while the Disaggregation Event records when a larger batch is divided into smaller parts for distribution. Additionally, the Transformation Event captures changes to a batch,

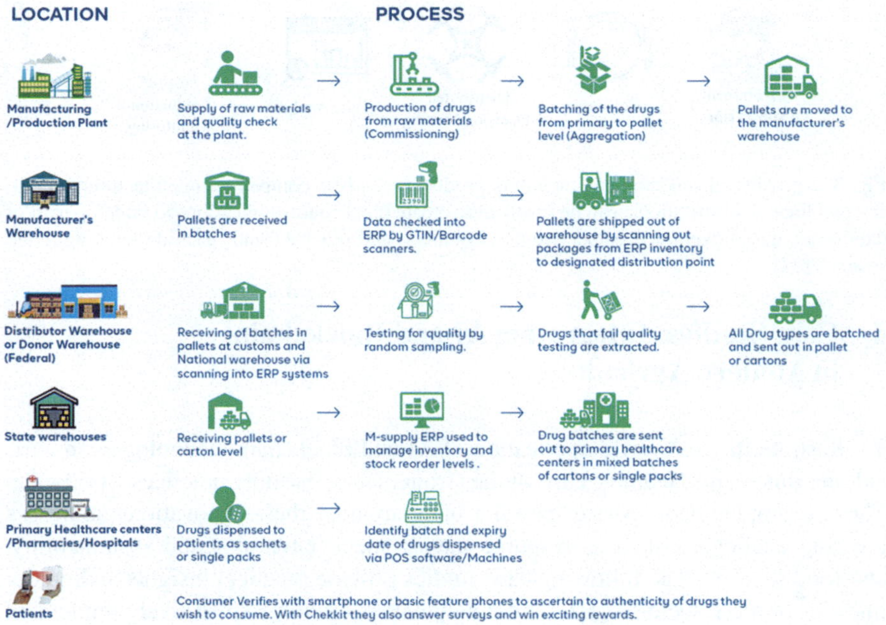

Fig. 4 Example of EPCIS-optimized process for the public PHARMA supply chain of anti-malaria drugs using GS1 EPCIS standards [27]

such as repackaging or relabeling, and the Inspection Event documents inspections by regulatory agencies. Finally, the Disposition Event tracks the disposal of pharmaceuticals, such as destruction due to expiration or recall, ensuring comprehensive oversight and traceability throughout the supply chain [27].

Like Farmer Connect, a blockchain platform in the coffee industry, IBM Food Trust leverages blockchain to facilitate real-time, end-to-end transparency. Food suppliers and distributors utilize the platform to trace products, ensuring compliance with stringent safety standards. Additionally, consumers can verify the authenticity and safety of their food by scanning a QR code to access essential product details, including its origin and supply chain journey.

Beyond blockchain, IBM Food Trust incorporates AI-driven predictive analytics to optimize logistics, reduce waste, and enhance production efficiency. AI tools process data from diverse sources, including market demand, weather patterns, and historical trends, enabling food suppliers to make informed decisions on resource allocation and production planning. These capabilities help predict potential disruptions and optimize the food supply chain [26].

IBM Food Trust addresses several pressing challenges in the global food industry, particularly in enhancing food safety, ensuring regulatory compliance (such as with FSMA Rule 204), and promoting sustainability. By fostering a digital ecosystem where stakeholders can share data seamlessly, the platform mitigates foodborne illnesses, reduces inefficiencies in the supply chain, and contributes to sustainability efforts by optimizing resource use and reducing waste [26].

The Food Traceability Final Rule (FSMA Section 204 which illustrated in Fig. 5) mandates record-keeping for stakeholders in the production, processing, or packaging of foods on the Food Traceability List (FTL), including leafy greens, melons, and peppers. Key Data Elements (KDE) linked to Critical Tracking Events (CTE)

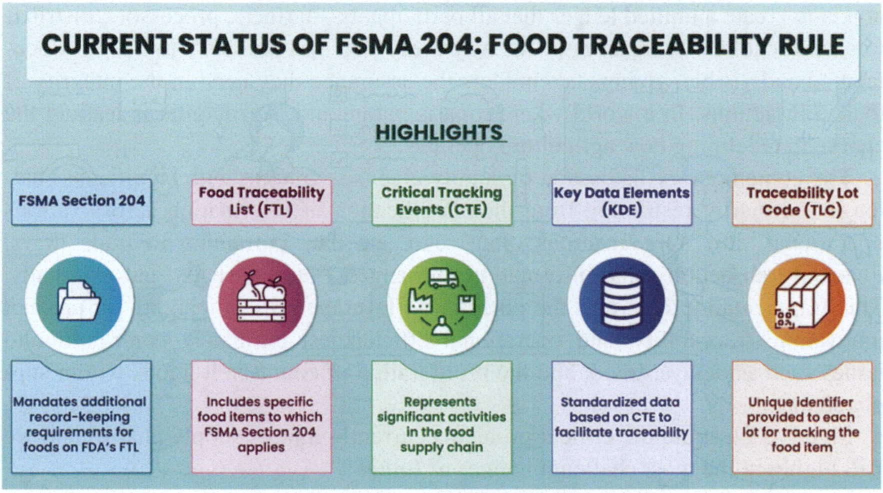

Fig. 5 Current status of FSMA 2024 food traceability rules [28]

enable swift identification of food safety incidents. Traceability Lot Codes (TLC) enhance the FDA's ability to monitor food movement and prevent recalls. Compliance with these regulations offers a competitive edge and improves food safety oversight. However, the food industry faces challenges related to managing traceability information, engaging stakeholders, and integrating technology. Addressing these challenges is essential for effective compliance and enhanced food safety [28].

IBM's partnership with iFoodDS further enhances the platform's traceability capabilities. This collaboration aims to expand the adoption of advanced safety standards and improve traceability for food suppliers on a global scale. By integrating blockchain and AI, IBM Food Trust positions itself as a transformative solution for future food supply chains, setting a new standard for safety, transparency, and sustainability.

This case study demonstrates the potential of integrating AI and blockchain to enhance transparency, safety, and operational efficiency within food supply chains. As the industry evolves, IBM Food Trust represents a significant benchmark for innovative practices in supply chain management, supporting broader global sustainability goals.

4.2 Case Study 2: AI-Driven Precision Farming and Blockchain for Sustainability: AgriDigital

Blockchain technology plays a pivotal role in this transformation. By offering a decentralized framework, it allows for real-time tracking of product movements and transactions, significantly boosting transparency. AgriDigital has embraced this technology to ensure comprehensive visibility across the entire supply chain. Their aim is to create a unified ledger that all participants—farmers, processors, distributors, and retailers—can access. This way every step of a product's journey is documented and visible, giving stakeholders the assurance they need in the integrity of their transactions. In a world where trust is paramount, AgriDigital is leading the charge in redefining how agriculture operates.

The contemporary landscape of supply chains is fraught with significant challenges, primarily restricting from the classification of information across various operational silos. Organizations often segregate data pertaining to trade flows, financial transactions, and information exchanges, both internally and externally. This fragmentation obstructs the potential to leverage critical insights for product innovation, traceability, and operational efficiencies, ultimately contributing to issues such as delays, fraud, and the proliferation of counterfeit goods within supply chains.

Figure 6 illustrate the critical problems currently impact supply chains, and we will highlight the most challenging ones as follows:

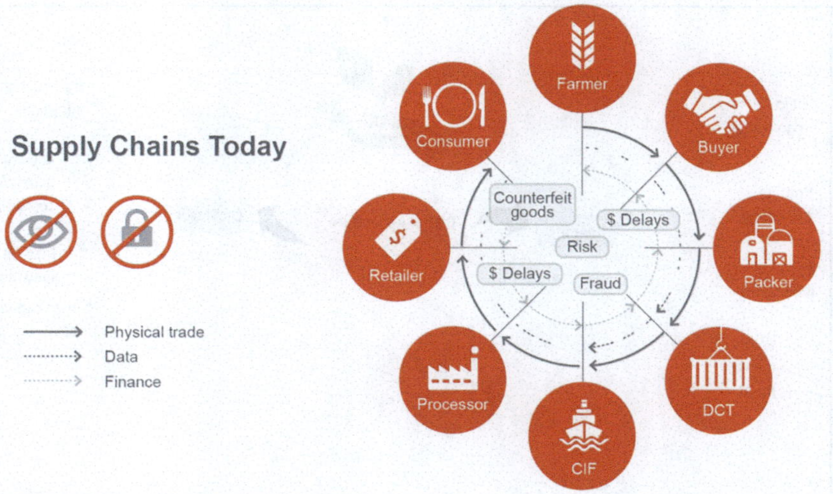

CIF = cost, insurance, and freight; DCT = digital control tower.

Fig. 6 Key challenges in modern supply chains [29]

- First, farmers frequently experience delays in receiving payment for their delivered commodities, with the Australian grains industry reporting an average payment processing time ranging from 2 to 5 weeks.
- Second, buyers encounter barriers to accessing financing that would facilitate timely payments to farmers. The complexity and inherent risks associated with commodity-backed finance dissuade many financial institutions from offering such support.
- Lastly, consumers grapple with a lack of trust and insufficient access to verifiable information regarding the status and quality of goods, which undermines their confidence in the supply chain. These interconnected issues underscore the urgent need for innovative solutions to enhance transparency and efficiency throughout the supply chain ecosystem [29].

AgriDigital exemplifies how blockchain technology is driving the digital transformation of agriculture by enhancing trust and transparency in supply chains. Combined with AI, drones, and sensors, these technologies create a "smart agriculture" ecosystem that improves efficiency, sustainability, and global food security. Together, they revolutionize farming processes, fostering innovation and new opportunities across value chains. Additionally, digital agriculture fosters international trade and improves animal welfare while driving innovation in value chains. Its potential to increase agricultural output benefits both society and the environment [30].

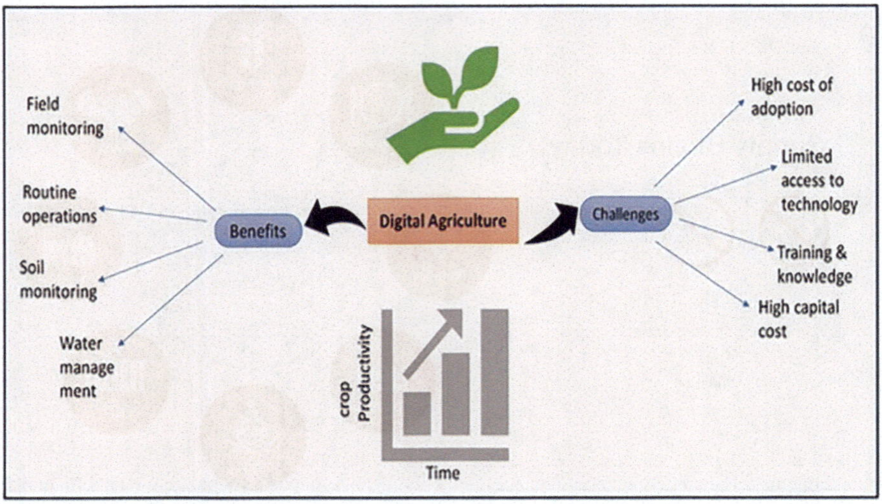

Fig. 7 A schematic representation of the benefits and challenges of digitization in agriculture [30]

Figure 7 outlines the benefits and challenges of digital agriculture and its impact on crop productivity over time.

The benefits include:

- **Field Monitoring:** Utilizes technology to observe and analyze field conditions, helping farmers make data-driven decisions.
- **Routine Operations:** Automates regular farming tasks, improving efficiency and reducing labor.
- **Soil Monitoring:** Involves assessing soil health and conditions, allowing for better management of nutrients and moisture levels.
- **Water Management:** Optimizes water usage, ensuring crops receive the right amount without waste, which is vital for sustainability.

The most important challenges:

- **High Cost of Adoption:** The initial investment in technology and equipment can be significant, deterring many farmers from transitioning to digital methods.
- **Limited Access to Technology:** Not all farmers have access to the latest technologies, which can create disparities in productivity and efficiency.
- **Training and Knowledge:** Farmers may lack the necessary skills or training to effectively use digital tools, which can limit their potential benefits.
- **High Capital Cost:** The ongoing costs associated with maintaining and upgrading technology can be prohibitive, especially for smaller operations.

4.3 Impact on Crop Productivity

Digital agriculture can be applied in various scenarios, with precision farming being a notable example. As shown in Fig. 7, precision farming is a crucial and sustainable strategy for enhancing crop production while improving the socio-economic conditions and livelihoods of farmers. This collection of methods, known as precision agriculture (PA), encompasses various agricultural research areas and is also referred to as variable rate technology, site-specific crop management, and prescription farming. As an innovative management approach, it utilizes georeferenced data to oversee agricultural systems, employs monitoring techniques, and integrates soil, plant, and climate factors. By utilizing computers, satellites, positioning systems, and remote sensing technologies, it provides detailed georeferenced information to facilitate better decision-making [30].

Figure 8 illustrates precision agriculture, integrating remote sensing, sensors, machine learning, and AI. Drones, equipped with sensors and microcontrollers, enable autonomous operation and have proven effective for chemical spraying, especially in hard-to-reach fields and cloudy conditions [31]. Their high image resolution offers an advantage over satellite sensors. Drones have been used in horticulture to spray artificial materials in targeted zones based on remote sensing data. They have also been fitted with low-volume sprayers for efficient pesticide application. As population growth and climate change drive the need for more efficient farming, drones will play an increasingly vital role in precision agriculture, from crop monitoring to soil analysis and spraying.

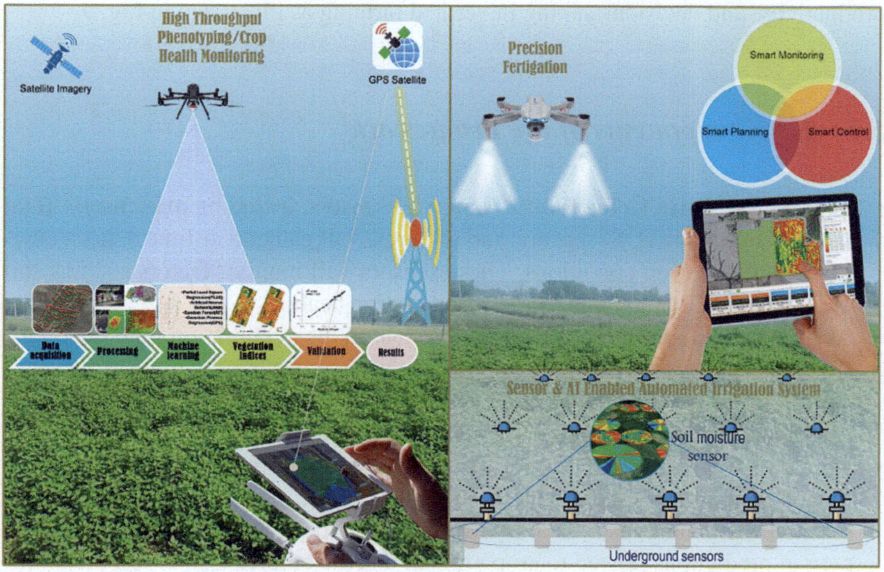

Fig. 8 Precision farming systems combined with drones, sensors, and artificial intelligence [30]

5 Agentic AI Workflow and Digital Twin for Optimizing Vertical Farming: A Proposed Framework

The growing worldwide populace, coupled with other urgent elements which include climate trade, a growing older population, and a shift of younger generations towards city and office jobs, has considerably elevated the call for meals manufacturing. This rising calls for, however, is met with exertions shortages, in addition exacerbating the demanding situations confronted by conventional agricultural systems. Advanced agricultural answers, which includes robotics, artificial intelligence (AI), and digital dual (DT) technologies, are rising as transformative equipment that promise to form the future of farming. These technologies offer better precision, performance, and sustainability, making them important in addressing the developing need for agricultural productivity and innovation. In this segment, we will provide an explanation for our ongoing studies thought, which specializes in developing a framework combining Agentic AI workflows and digital twins technology to orchestrate and optimize vertical farming.

5.1 Agentic AI Workflows in Vertical Farming

Our proposed Digital Twin and Agentic AI Workflow framework is designed to create an incorporated approach for optimizing vertical farming practices using advanced technology, along with machine learning, AI, and real-time records. The key components of this framework work collaboratively to enhance sustainability, productivity, and decision-making in vertical farming as illustrated in Fig. 9.

5.2 Key Components of the Framework

- **Sensors and Data Collection:** A variety of sensors, including pH sensors, Total Dissolved Solids (TDS) sensors, and cameras, will gather real-time data on environmental and crop conditions. These sensors will monitor factors such as soil moisture, temperature, nutrient levels, and crop health.
- **Actuators and Automated Systems:** Actuators like calibration pumps, robots, sprayers, and fertilizer applicators will execute actions based on the recommendations of AI agents. This includes automated actions such as adjusting the level of nutrient, irrigating crops, or harvesting during the optimal conditions.
- **Machine Learning and Model Training:** By using online machine learning models we continuously update the system's predictive capabilities based on the recent drifts in the sensor data. By using these models, we can improve the accuracy of predictions related to crop growth, disease outbreaks, and resource utilization, ensuring that the system remains responsive to any changing conditions.

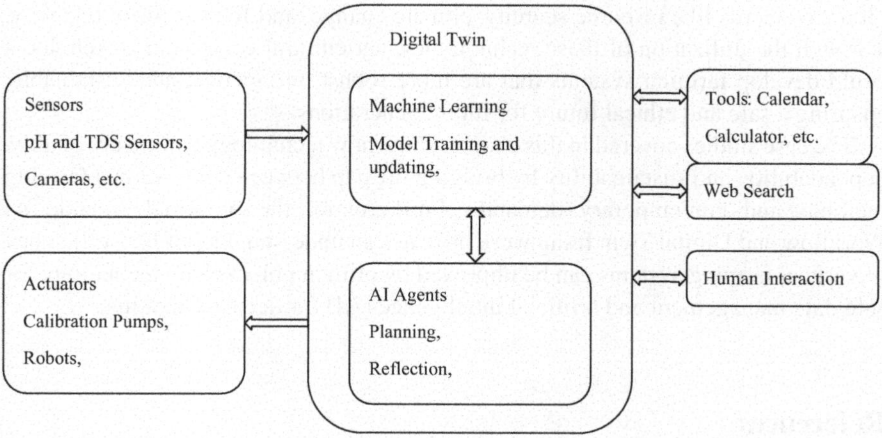

Fig. 9 Agentic AI and digital twin framework

- **AI Agents:** The agents will optimize resource utilization, decision-making, and action planning in a multi-agent environment. By communicating with sensors, actuators, external data sources and each other, these agents will enable real-time, intelligent farm management.
- **Human Interaction:** Farmers and experts can interact with the system by inputting data, reviewing recommendations, and validating the decisions made by the AI agents.
- **Web Search and External Data Integration:** The AI agents will have the capability to access real-time data from the internet, such as weather forecasts, market trends, and academic studies. This will allow the system to refine its knowledge based on the latest available information.
- **Tools for Planning and Resource Management:** The system will provide various tools, including calculators and calendars, to help to plan and manage resources. These tools will enable farmers to make informed decisions regarding planting schedules, resource allocation, and best market timing.

6 Conclusion

With the help of blockchain technology and artificial intelligence, smart farming is revolutionizing the agricultural industry and providing innovative solutions to its major issues. AI-driven solutions enable farmers forecasting and automation capabilities to optimize operations and resource use, facilitating precision, efficiency, and sustainability. At the same time, blockchain technology preserves accountability, transparency, and trust throughout the supply chain for agricultural products, promoting a more robust and ethical food system. The adoption of blockchain technology and artificial intelligence in agriculture is going to increase in future as global concerns like resource scarcity, climate change, and food security intensify. Through the utilization of these technologies, agricultural ecosystem stakeholders could develop farming systems that are more robust, productive, and sustainable, ensuring a safe and ethical future for future generations.

The case studies covered in this chapter show how technology can ensure quality, dependability, and sustainability by bridging the gap between conventional farming methods and contemporary demands. Furthermore, the proposed Agentic AI Workflow and Digital Twin framework provides an understanding of how efficiency of vertical farming systems can be improved by utilizing blockchain technology for safe data management and artificial intelligence (AI) for decision-making.

References

1. World Population Prospects. (2019). *World population projected to reach 9.8 billion in 2050 and 11.2 billion in 2100*. United Nations. Retrieved December 12, 2024, from https://www.un.org/en/desa/world-population-projected-reach-98-billion-2050-and-112-billion-2100
2. Idrees, S. M., Nowostawski, M., Jameel, R., & Mourya, A. K. (2021a). Security aspects of blockchain technology intended for industrial applications. *Electronics, 10*, 951. https://doi.org/10.3390/electronics10080951
3. Kamilaris, A., Kartakou llis, A., & Prenafeta-Boldú, F. X. (2017). A review on the practice of big data analysis in agriculture. *Computers and Electronics in Agriculture, 143*, 23–37.
4. Liakos, K. G., Busato, P., Moshou, D., Pearson, S., & Bochtis, D. (2018). Machine learning in agriculture: A review. *Sensors, 18*(8), 2674.
5. Meshram, V., Patil, K., Meshram, V., Hanchate, D., & Ramkteke, S. D. (2021). Machine learning in horticulture domain: A state-of-art survey. *Artificial Intelligence in the Life Sciences, 1*, 100010.
6. Kienzle, J. (2013). Precision agriculture for smallholder farmers. *Agriculture for Development, 12*, 12–15.
7. Shamshiri, R. R., Kalantari, F., Ting, K. C., Thorp, K. R., Hameed, I. A., & Weltzien, C. (2018). Advances in greenhouse automation and controlled environment agriculture: A transition to plant factories and urban agriculture. *International Journal of Agricultural & Biological Engineering, 11*(1), 1–22.
8. Ryan, M. (2023). *The social and ethical impacts of artificial intelligence in agriculture: Mapping the agricultural AI literature* (Vol. 38, p. 2473). AI & Society.
9. Tantalaki, N., Koukouvinis, E., & Vassilopoulou, M. (2019). The role of artificial intelligence in agriculture: A review. *Journal of Agricultural Informatics, 10*(2), 1–9.
10. Banerjee, A., Bandyopadhyay, T., & Acharya, P. (2013). Data analytics: Hyped up aspirations or true potential? *Vikalpa, 38*, 1–12.
11. Gardezi, M., Joshi, B., Rizzo, D., Ryan, M., Prutzer, E., Brugler, S., & Dadkhah, A. (2024). Artificial intelligence in farming: Challenges and opportunities for building trust. *Agronomy Journal, 116*, 1217. https://doi.org/10.1002/agj2.21353
12. Rossel, R. A. V., & Bouma, J. (2016). Soil sensing: A new paradigm for agriculture. *Agricultural Systems, 148*, 71–74.
13. Smith, M. J. (2018). Getting value from artificial intelligence in agriculture. *Animal Production Science, 60*(1), 46–54.
14. Koliba, C., & Spett, E. (2023). What a democratically anchored public administrator needs to understand about artificial intelligence and strategic management. In C. Greve & T. Ysa (Eds.), *Handbook on strategic public management*. Edward Elgar Publishing.

15. Bermeo-Almeida, O., Cardenas-Rodriguez, M., Samaniego-Cobo, T., Ferruzola-Gómez, E., Cabezas-Cabezas, R., & Bazán-Vera, W. (2018). Blockchain in agriculture: A systematic literature review. In R. Valencia-García, G. Alcaraz-Mármol, J. Del Cioppo-Morstadt, N. Vera-Lucio, & M. Bucaram-Leverone (Eds.), *Technologies and innovation. CITI 2018. Communications in computer and information science* (Vol. 883). Springer. https://doi.org/10.1007/978-3-030-00940-3_4
16. Idrees, S. M., Aijaz, I., Jameel, R., & Nowostawski, M. (2021b). Exploring the blockchain technology: Issues, applications and research potential. *International Journal of Online and Biomedical Engineering (iJOE), 17*(07), 48–69. https://doi.org/10.3991/ijoe.v17i07.22803
17. Menon, S., & Jain, K. (2024). Blockchain technology for transparency in agri-food supply chain: Use cases, limitations, and future directions. *IEEE Transactions on Engineering Management, 71*, 106–120.
18. Ge, L., Brewster, C., Spek, J., & van Diepen, F. (2021). Blockchain for agriculture and food: Findings from the pilot study in The Netherlands. *Journal of Agricultural Informatics, 12*(1), 45–56.
19. Lin, W., Zhu, J., & Chen, S. (2020). Blockchain technology for enhancing food supply chain traceability and transparency. *Journal of Cleaner Production, 257*, 120543.
20. El Mane, A., Tatane, K., Chihab, Y., & Mane, K. (2024). Transforming agricultural supply chains: Leveraging blockchain-enabled Java smart contracts and IoT integration. *ICT Express, 10*(3), 650–672.
21. Tan, B., Yan, J., Chen, S., & Liu, X. (2018). The impact of blockchain on food supply chain: The case of Walmart. In M. Qiu (Ed.), *Smart blockchain* (Vol. 11373, pp. 3–18). Springer.
22. Kamilaris, A., & Prenafeta-Boldú, F. X. (2022). The rise of blockchain in agriculture: A review and outlook. *Computers and Electronics in Agriculture, 193*, 106658.
23. Wang, R., Qian, H., & Wang, Y. (2023). Blockchain-based agricultural commodity trading platforms: Opportunities and challenges. *Agricultural Systems, 201*, 103484.
24. Ehsan, I., Khalid, M. I., Ricci, L., Iqbal, J., Alabrah, A., & Ullah, S. S. (2022). A conceptual model for blockchain-based agriculture food supply chain system. *Scientific Programming, 2022*, 358354. https://doi.org/10.1155/2022/7358354
25. Zhao, M. J., Shi, R., & Yao, L. Y. (2022). Analysis on the goals and paths of carbon neutral agriculture in China. *Issues in Agricultural Economy, 9*, 24–34.
26. Wiliot. (2023). *IBM Food Trust: Transforming the future of food safety and supply chains*. Retrieved October 4, 2024, from https://www.wiliot.com/podcasts/ibm-food-trust-transforming-the-future-of-food-safety-and-supply-chains
27. Chekkit Technologies. (2023). *What is the GS1 EPCIS standard in supply chain traceability?* Retrieved October 8, 2024, from https://chekkitapp.com/blog/what-is-the-gs1-epcis-standard-in-supply-chain-traceability
28. Food Safety Select. (2024). *Global food chain traceability and the current status of FSMA's food traceability rule*. Retrieved October 15, 2024, from https://www.foodsafetyselect.com/business-news/global-food-chain-traceability-and-the-current-status-of-fsma%E2%80%99s-food-traceability-rule
29. Asian Development Bank (ADB). (2021). *Transforming agricultural supply chains using blockchain*. Retrieved from October 25, 2024, from https://development.asia/explainer/transforming-agricultural-supply-chains-using-blockchain
30. Balyan, S., Jangir, H., Tripathi, S. N., Tripathi, A., Jhang, T., & Pandey, P. (2024). Seeding a sustainable future: Navigating the digital horizon of smart agriculture. *Sustainability, 16*(2), 123–145.
31. Spoorthi, S., Shadaksharappa, B., Suraj, S., & Manasa, V. K. (2017). Freyr drone: Pesticide/fertilizers spraying drone—An agricultural approach. In *Proceedings of the 2nd International Conference on Computing and Communications Technologies (ICCCT), Chennai, India, 23–24 February 2017.*

Revolutionizing Education: The Role of AI and Blockchain

Khyati Marwah, Jaweria Izhar, Rashi Singh, and Roshan Jameel

1 Introduction

In recent years, technological advancements have profoundly impacted various sectors, with education undergoing significant transformation. Among the most influential innovations are Artificial Intelligence (AI) and Blockchain, which hold the potential to address persistent challenges such as inequitable access, administrative inefficiencies, and the demand for secure credentialing systems. AI drives educational progress by enabling personalized learning experiences tailored to individual student needs, employing adaptive assessments that dynamically adjust to skill levels, and deploying intelligent systems capable of accommodating diverse learning styles. For instance, AI-powered platforms like adaptive tutoring systems can identify gaps in a student's understanding and offer targeted resources, akin to having a dedicated personal tutor.

Similarly, Blockchain technology provides secure and decentralized solutions for credential management, ensuring trust, transparency, and the integrity of academic records [1]. An analogy might be a digital notary system that verifies the authenticity of credentials, making them tamper-proof and universally accessible, particularly beneficial in global academic and professional contexts. Together, AI and Blockchain are reshaping the educational landscape, aligning it with the demands of an increasingly interconnected and rapidly evolving global society. These technologies not only enhance efficiency and accessibility but also pave the way for more inclusive and future-ready educational systems.

K. Marwah · J. Izhar (✉) · R. Singh
DeMont Institute of Management and Technology, Dubai, UAE
e-mail: Jaweria.i@demont.ac.ae

R. Jameel
Westford University College, Sharjah, UAE

S. M. Idrees et al. (eds.), *Blockchain and Machine Learning Innovations*, Signals and Communication Technology,
https://doi.org/10.1007/978-3-031-88095-7_12

The chapter is structured to provide a comprehensive exploration of how Artificial Intelligence (AI) and Blockchain can revolutionize education, offering both practical insights and strategic guidance. Section 1 introduces the foundational concepts of AI and Blockchain, emphasizing their defining features and key applications within educational contexts. Followed by Sect. 2 that discusses the literature work done in the field. Section 3 provides a comparative analysis of traditional education methods versus modern approaches powered by AI and Blockchain. Section 4 presents a real-world case study that illustrates the practical integration of AI and Blockchain into educational system, followed by Sect. 5 that discusses the technical challenges and their solutions for adopting AI-Blockchain-based educational systems. Finally, Sect. 6 the Conclusion synthesizes the chapter's key findings, offering actionable recommendations for future strategies aimed at leveraging AI and Blockchain in education.

By presenting these arguments in a structured manner, the chapter offers a holistic understanding of AI and Blockchain's role in enhancing educational systems. It not only highlights their potential to drive innovation but also underscores the importance of addressing critical challenges to unlock their full benefits.

1.1 Role of AI and Blockchain in Education

Artificial Intelligence (AI) refers to the simulation of human intelligence by computer systems, encompassing functions such as learning, reasoning, and problem-solving. When applied to education, AI leverages advanced technologies to support teachers and students in achieving optimal learning outcomes [2]. For example, AI-driven tools such as Knewton Alta [3], DreamBox Learning [4] can analyze student performance data to create personalized learning pathways, adapting instructional methods to individual needs and fostering more effective engagement.

Blockchain, on the other hand, is a distributed, tamper-proof ledger designed to record transactions and track assets securely and efficiently. Assets can be tangible, such as property, vehicles, or currency, or intangible, such as intellectual property, patents, and copyrights. In educational contexts, Blockchain technology facilitates the secure and transparent management of credentials and records. By providing an immutable, time-stamped catalogue of events in real time, Blockchain ensures data integrity, reduces administrative inefficiencies, and enhances trust in academic processes [5].

Together, AI and Blockchain represent transformative technologies that address both pedagogical and administrative challenges in education. AI enhances learning experiences by personalizing instruction and automating repetitive tasks, while Blockchain ensures the secure verification and transfer of educational credentials, paving the way for more equitable and efficient systems.

1.2 *An Overview of AI's Benefits and Challenges in Education*

Artificial Intelligence (AI) significantly enhances the learning process by enabling personalized learning experiences and providing immediate feedback. Defined as the capability of computer systems to perform tasks traditionally requiring human judgment, estimation, and guidance [2], AI has emerged as a transformative force in education. Its applications in higher education include robot writers, virtual experiences, and intelligent tutoring systems (ITSs). While the prospect of human-like robot teachers may remain distant, current AI technologies already offer practical tools for students, educators, and institutions to improve educational experience.

1.2.1 Benefits of AI in Education

AI offers several promising advantages in education, including:

- **Personalized Learning**: AI adapts instructional content and pace to meet individual learner needs, ensuring students receive information in a manner that maximizes comprehension and retention. For instance, adaptive learning platforms can tailor exercises and resources based on a learner's progress and challenges.
- **Enhanced Learning Experiences**: By integrating multimedia and interactive elements, AI fosters a more engaging and immersive educational environment. Features such as graphics, animations, and interactive exercises make learning visually stimulating and appealing to diverse learning styles.
- **Increased Accessibility**: AI-driven technologies create inclusive learning opportunities for students with diverse abilities. Tools like speech-to-text applications, personalized learning devices, and virtual assistants help overcome traditional barriers to education, ensuring broader access for all learners.
- **Streamlined Administrative Tasks**: AI automates repetitive administrative functions, such as grading assignments and scheduling, reducing educators' workload. This optimization allows teachers to focus more on instruction and student engagement.
- **Real-Time Feedback**: AI systems provide instantaneous feedback on student performance, helping identify areas that require improvement. This immediacy supports smoother learning.

1.2.2 Challenges of AI in Education

Despite its numerous benefits, implementing AI in education presents significant challenges that must be addressed to ensure its effective and ethical use. These challenges include:

- **Ethical Concerns**: Issues such as algorithmic bias and potential racism in AI systems can perpetuate inequality if not properly mitigated through robust oversight and inclusive design practices. Ethical dilemmas also arise around data privacy and security, as AI systems often require extensive data collection. Ensuring the responsible and ethical use of AI is essential to build trust and protect the rights of individuals within educational institutions [2].
- **Data Privacy and Security**: The reliance on large volumes of sensitive student data raises critical concerns about privacy and security. Institutions must implement stringent measures to safeguard this information, as breaches can have severe consequences for both learners and educators.
- **Impact on the Workforce**: The automation of tasks traditionally performed by educators, such as grading and administrative processes, can lead to concerns about job displacement and changing workforce dynamics. Balancing the use of AI with the preservation of meaningful human roles in education is essential to address these apprehensions.
- **Implementation Barriers**: The integration of AI into educational systems faces practical and financial challenges. Schools and universities often require significant investment in infrastructure, training, and technical support to facilitate smooth implementation. Resource disparities between institutions further exacerbate this issue.
- **Technology Dependence**: Over-reliance on AI technologies could diminish critical thinking and social interaction skills among students. As AI continues to permeate educational practices, it is crucial to maintain a balance between technological innovation and traditional learning methods to preserve the human aspect of education [2].
- **Access and Equity**: While AI has the potential to improve access to education, it can also exacerbate inequalities. The digital divide, stemming from unequal access to technological resources, remains a pressing concern. Ensuring equitable access to AI-enhanced learning tools is vital to avoid deepening disparities among students.

With careful planning and strategic implementation [2], these challenges can be effectively addressed. Institutions must prioritize equity, inclusivity, and ethical considerations to fully harness AI's transformative potential while mitigating its associated risks.

1.3 An Overview of Blockchain's Benefits and Challenges in Education

Building on the discussion of AI, Blockchain emerges as a complementary technology that addresses critical challenges in education, including security, transparency, and credential verification. Blockchain's unique characteristics—decentralization, immutability, and transparency—enable innovative applications that can transform various aspects of educational systems [6].

For example, academic records such as transcripts, degrees, and test results can be securely stored and verified on blockchain networks, ensuring their authenticity, and reducing the risk of fraud. Additionally, smart contracts can automate agreements between professors, university officials, and students, such as setting assignment deadlines or managing student loan payments. Tokenization strategies, such as awarding cryptocurrency for academic achievements, have also demonstrated effectiveness in motivating students and fostering accountability.

Although still in its early stages, blockchain technology shows immense potential to increase transparency, enhance accountability, and incentivize participation and learning. Educators, as noted by Anderson et al. [2], are looking toward technology as a vital tool for modernizing education.

1.3.1 Key Applications of Blockchain in Education

- **Credential Verification**: Blockchain can securely store and verify academic credentials, such as degrees, certificates, and transcripts. This ensures the authenticity of achievements while reducing the prevalence of credential fraud. Employers and institutions can quickly verify qualifications, saving time, and resources.
- **Transparency and Tamper-Proof Records**: With its distributed ledger system, blockchain ensures that data cannot be altered or deleted once recorded. This feature is invaluable for maintaining the integrity of student records and safeguarding academic achievements against unauthorized tampering.
- **Decentralized Learning Platforms**: Blockchain supports peer-to-peer learning platforms where students and educators can interact directly without intermediaries. Such platforms enable personalized and efficient learning experiences while empowering users with greater control over their data and interactions.
- **Micropayments for Educational Content**: Blockchain allows cryptocurrency-based micropayments, making it possible for students to purchase microlearning modules or specific course content. This model promotes accessibility by enabling learners to pay only for the content they need.
- **Smart Contracts for Administrative Processes**: Smart contracts automate routine administrative tasks, such as enrollment, course registration, and grade tracking. By minimizing paperwork and reducing errors, these self-executing contracts improve efficiency and reliability in managing academic processes.

Blockchain technology holds transformative potential in education by addressing longstanding challenges and introducing innovative solutions [7]. From securing academic records to decentralizing learning platforms, blockchain enhances transparency, accountability, and efficiency. As institutions increasingly explore its applications, blockchain is poised to play a pivotal role in modernizing education systems and fostering an equitable, technology-driven learning environment.

1.3.2 Limitations and Challenges of Blockchain Technology in Education

While blockchain technology offers significant benefits, its adoption in education is accompanied by notable limitations and challenges that must be addressed to ensure effective implementation. These include:

- **Transaction Speed**: Blockchain networks often face scalability issues, leading to slower transaction processing speeds as the network expands. For example, widely used platforms like Bitcoin and Ethereum experience delays due to their limited throughput capabilities. This could pose challenges for educational applications that demand high-speed data transactions, such as real-time credential verification or dynamic learning platforms.
- **Energy Consumption**: Many blockchain networks rely on proof-of-work (PoW) consensus algorithms, which are energy-intensive. Mining activities, involving complex mathematical computations, consume vast amounts of energy and are environmentally unsustainable. For educational institutions aiming to adopt blockchain, the environmental impact of PoW systems might deter large-scale implementation.
- **Data Storage Limitations**: As blockchain networks grow, the size of the ledger increases, leading to significant storage requirements. Storing large volumes of data on every node becomes increasingly challenging, potentially limiting participation in the network. For education systems managing extensive student records and credentials, this could become a major obstacle.
- **Lack of Regulation and Standards**: The legal and regulatory framework governing blockchain and cryptocurrency remains underdeveloped. The absence of clear guidelines creates ambiguity and uncertainty, making organizations and individuals hesitant to adopt blockchain solutions. For instance, educational institutions may be cautious about investing in blockchain technologies without assurance of compliance and legal protections.
- **Irreversibility of Transactions**: Blockchain's immutability, while a strength in ensuring data integrity, becomes a liability in cases of error or fraud. Once a transaction is recorded on the blockchain, it cannot be altered or reversed, complicating the correction of mistakes or unauthorized activities. In an educational setting, this limitation could result in challenges, such as resolving disputes over inaccurate credential records.

The limitations of blockchain, including scalability issues, high energy consumption, storage challenges, regulatory uncertainties, and the irreversible nature of transactions, highlight the need for careful consideration when implementing this technology in education. Addressing these challenges through innovative solutions—such as adopting energy-efficient consensus mechanisms, enhancing scalability, and developing clear regulatory frameworks—will be critical for leveraging blockchain's full potential in transforming educational systems.

1.4 Evaluation of the Current Challenges in Education Domain

This paper provides an in-depth review of the critical challenges currently plaguing global educational systems, emphasizing several key areas. These challenges include:

- **Technology Integration**: Despite the rapid growth of digital tools in education, integrating these technologies into the teaching and learning process remains a major struggle. Disparities in access to technology and digital skills among students and educators persist [8]. Furthermore, the sudden shift to remote learning during the COVID-19 pandemic revealed infrastructural limitations, including inequities in device availability and teacher preparedness for online instruction [9].
- **Teacher Adaptation**: Educators face significant challenges in adjusting to new teaching methods, learning technologies, and the pressure of managing various devices. There is a clear need for continuous professional development programs to enhance teachers' digital literacy and pedagogical skills, enabling them to effectively use technology in their teaching.
- **Student Engagement and Mental Health**: Engagement in virtual or hybrid classes has been a challenge, impacting student motivation and academic performance. Furthermore, the effects of academic failure on student mental health have emerged as a pressing issue, necessitating stronger campus support services to address students' emotional and psychological needs.
- **Assessment Methods**: Traditional modes of assessment are being questioned for their ability to accurately measure twenty-first-century skills. This requires a rethinking of evaluation strategies to better align with the evolving demands of education and the workforce.
- **Equity and Inclusion**: Ongoing issues of equity and inclusion highlight the need for concerted efforts to provide equal educational opportunities for students from diverse backgrounds, ensuring that no group is left behind.
- **Curriculum Alignment**: The need for curriculum updates to better align with changing workforce demands is crucial. Educational institutions must ensure that the curriculum develops relevant skills for students, preparing them for modern job markets.

This review adds valuable insights to the ongoing discourse on the challenges facing education today, offering lessons that can guide future policy decisions and transformative interventions aimed at improving the quality of education worldwide.

2 Literature Review

The literature review provides a systematic evaluation of AI and Blockchain applications in education, highlighting their current state, benefits, challenges, and potential future uses. While AI offers tools for personalized learning and efficiency, Blockchain ensures security, transparency, and integrity in academic processes. Their combined application can redefine educational paradigms, fostering a future marked by innovation, inclusivity, and accessibility. This foundation sets the stage for exploring their broader implications and addressing the challenges of their integration into educational ecosystems.

AI has gained traction in education through innovations such as intelligent tutoring systems, adaptive assessments, and automated administrative tasks. For instance, natural language processing and machine learning algorithms analyze extensive datasets to offer personalized educational experiences. Anderson et al. [2] emphasize that AI-powered platforms actively engage students, enhancing learning outcomes. Tools like AI grading systems facilitate more meaningful teacher-student interactions, allowing educators to focus on qualitative aspects of teaching [10]. On the other hand, Blockchain's decentralized and immutable ledger system addresses key challenges in record-keeping, credential verification, and identity management. Johnson et al. [11] highlight that Blockchain ensures the integrity of academic credentials by reducing fraud and providing a secure platform for record storage and verification. Decentralized identity management systems can streamline enrolment processes while preventing identity-related fraud.

The integration of AI and Blockchain offers a powerful combination of transparency, efficiency, and security. Research by Chen et al. [12] demonstrates that Blockchain secures the provenance of data used by AI algorithms, ensuring the reliability of AI-driven educational insights. Smart contracts enabled by Blockchain further automate and secure administrative processes, from managing assignment deadlines to facilitating financial transactions. These applications enhance trust, accountability, and efficiency across educational systems [13].

A recent study delves into the opportunities and challenges surrounding the application of Blockchain technologies in education [5]. It highlights key applications, particularly the digitalization and decentralization of educational certifications, while also discussing significant challenges such as data protection laws and scalability. The study emphasizes the relative immaturity of Blockchain innovations in education and the cautious perceptions held by governance bodies within educational institutions, which hinder rapid adoption. Despite a growing body of literature on the topic, few studies have focused on specific blockchain-based projects within higher education. As Anderson et al. [2] says there remains a need for a systematic review that explores the integration of AI and Blockchain protocols in higher education projects. This review would contribute to improving the application of these technologies by analyzing the current features and services available in the field.

Chaka [14] offers a comprehensive review that synthesizes the applications, prospects, and challenges of AI, robotics, and blockchain in higher education institutions. The findings emphasize the significant role of AI-driven chatbots in learning, the exploratory use of robotics, and various applications of blockchain technology, such as digital grading systems and real-time contract management. The study further examines blockchain's role in credentialing within the medical field, where it helps streamline the process by ensuring competency documents are reported in a unified format, thereby enhancing the integrity of credential distribution. Johnson et al. [15] provides an extensive review of blockchain applications in education, focusing on advanced applications in learning record-keeping and the establishment of a decentralized educational ecosystem. The review discusses the technical challenges faced by blockchain systems, including scalability issues, and suggests off-chain solutions like sidechains to address these challenges without compromising security and decentralization. Additionally, privacy-preserving AI techniques, such as homomorphic encryption, are discussed to ensure the security of sensitive data while enabling robust analytics.

Moizumder et al. [16] introduces the concept of the metaverse, examining its integration into the educational sector. They explore how the convergence of AI and the Internet of Things (IoT) facilitates the creation of digital worlds that provide flexible and secure engagement in social and educational activities. This paper discusses the historical context, current features, and prospects of the metaverse, highlighting its potential to revolutionize the educational experience. [17] explore the rapid growth of e-learning systems and propose blockchain-based IoT architecture to secure online education. This framework enhances distance learning programs, improving student learning experiences and outcomes. The integration of blockchain technology ensures the security and reliability of digital learning platforms, enabling more efficient and effective educational delivery.

The review of existing literature underscores the transformative potential of AI, Blockchain, and related technologies in education. While significant progress has been made in exploring their applications, challenges such as scalability, data privacy, and institutional hesitancy remain. Nevertheless, the integration of these technologies promises to enhance security, transparency, and personalization in education, offering a path toward more efficient, equitable, and accessible learning environments. This body of research lays a solid foundation for further exploration into the practical applications of AI and Blockchain in higher education, with the aim of improving educational outcomes globally.

3 The Evolution of Education: Traditional vs. Tech-Driven

The contemporary educational landscape faces persistent challenges, including outdated systems, inequitable access to quality learning resources, and non-transparent academic credential verification processes. A comparison is done in Table 1 below. These issues necessitate the adoption of integrated, state-of-the-art solutions like

Table 1 Detailed comparative analysis: traditional education vs. AI and blockchain education

Aspect	Traditional education	AI and blockchain education
Accessibility	Limited by physical location, resources, and infrastructure	AI-powered platforms offer online courses, breaking geographical barriers. Blockchain ensures decentralized and secure access to educational resources and credentials
	Scenario: Students in remote areas may lack access to quality schools and teachers	**Scenario**: A student in a rural area accesses AI-driven platforms like Coursera, receiving personalized lessons and verified certificates via Blockchain
Personalization	Difficult to cater to individual learning styles and paces due to standardized methods	AI algorithms analyze student performance and preferences, delivering adaptive content tailored to individual needs. Blockchain stores secure academic records for personalized progress tracking
	Scenario: Teachers use the same curriculum for all students, regardless of their strengths or weaknesses	**Scenario**: An AI tutor adjusts difficulty levels in real time for a struggling math student, while Blockchain stores their achievements and personalized recommendations
Cost	High costs include tuition, accommodation, and transportation fees	Online education platforms reduce tuition costs. Blockchain eliminates intermediaries, lowering administrative expenses
	Scenario: A university degree costs thousands of dollars annually, including dorm fees and books	**Scenario**: A student completes a specialized AI course for a fraction of the cost, receiving a Blockchain-verified certificate without administrative overheads
Flexibility	Fixed schedules and rigid structures limit opportunities for working professionals	AI-driven platforms enable learning anytime, anywhere. Blockchain maintains records that are accessible globally
	Scenario: A full-time worker struggles to attend night classes at a local university	**Scenario**: The same worker learns online during free hours, accessing AI-curated modules and earning certifications stored securely on Blockchain
Quality of content	Relies on the expertise of instructors and outdated textbooks	AI curates and generates up-to-date, high-quality content. Blockchain ensures authenticity and immutability of materials and certifications
	Scenario: Students in a traditional setup use outdated materials provided by schools	**Scenario**: Learner's access cutting-edge AI-generated content with Blockchain verifying its originality and academic credibility
Assessment and feedback	Assessments are subjective and delayed, limiting timely interventions	AI automates assessments, offering instant, data-driven feedback. Blockchain securely stores assessment results and credentials
	Scenario: A student waits weeks for results and receives generic feedback	**Scenario**: The same student receives instant, AI-generated feedback on an essay, while Blockchain ensures the security and accessibility of their scores for future reference

Artificial Intelligence (AI) and Blockchain. AI can facilitate personalized learning, tailoring educational content and pace to individual student needs while enhancing the efficiency of educational processes. Simultaneously, Blockchain technology offers robust solutions for credentialing and authentication, ensuring that academic records are secure, immutable, and globally accessible. However, integrating these technologies comes with technical, ethical, and regulatory complexities. Addressing these requires innovative approaches that balance technological advancements with social and ethical considerations.

4 Case Study: Learn Sync: Innovating with AI and Blockchain for Learning

4.1 AI in Personalized Learning

Learn Sync EdTech Solutions utilizes AI to deliver highly personalized and adaptive learning experiences for students. The platform employs machine learning algorithms to analyze individual learning patterns, strengths, and weaknesses in real time. Based on this analysis, it generates customized learning paths and content recommendations tailored to each student's needs.

In a 2020 implementation across a school district, the AI-driven platform analyzed students' learning styles and continuously monitored their progress. The platform featured adaptive learning modules that dynamically adjusted the difficulty levels, providing students with challenges suited to their abilities, thereby enhancing personalized engagement. Teachers also benefited from real-time analytics, which allowed them to identify struggling students and provide targeted interventions. This approach led to significant improvements, including a 30% increase in student engagement and a 20% boost in retention rates compared to traditional teaching methods. Furthermore, the automation of administrative tasks reduced teachers' workloads, providing more time for mentorship and student interactions.

4.2 Blockchain in Personalized Learning

In response to concerns about the authenticity and integrity of digital credentials, Learn Sync integrated blockchain technology into its platform in 2021. The blockchain-enabled system provides a secure and decentralized ledger for issuing and verifying educational credentials.

This integration allows the platform to store digital credentials such as course completions and academic achievements securely on the blockchain, ensuring data integrity and transparency. The blockchain system also offers immediate verifiability, allowing educational institutions and employers to authenticate credentials quickly and without the risk of tampering. This innovation gave students more control over their academic records, enabling seamless sharing of credentials with universities and potential employers. Blockchain system resulted in a 40% reduction in the time required for credential verification by educational institutions, significantly enhancing administrative efficiency and reducing administrative burdens.

As depicted in Fig. 1, AI has the potential to reduce teacher burnout by automating routine administrative tasks [18].[1] This benefit was clearly seen in Learn Sync's implementation, which automated processes such as grading, attendance tracking, and reporting. By doing so, the platform reduced the administrative workload for teachers, allowing them to spend more time focusing on student development and personalized instruction.

In conclusion, Learn Sync's experience underscores the transformative potential of integrating AI and blockchain into education, enhancing personalized learning, improving credential verification, and reducing the administrative burden on teachers. This case study provides valuable insights for future efforts to leverage these technologies in the education sector.

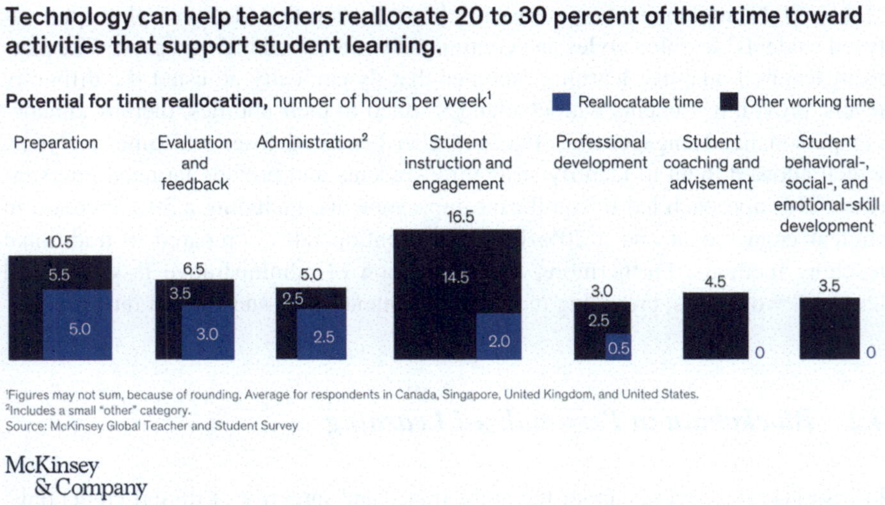

Fig. 1 How AI can accelerate students' holistic development and make teaching more fulfilling

---

[1] Figure 2: Kopp (2023) [18].

5 Technical Challenges and Solutions in AI and Blockchain in Education

While the integration of AI and blockchain into education offers immense potential, it also brings about several technical challenges. Fortunately, innovative solutions are emerging to address these challenges and enhance the effectiveness of these technologies in educational settings. Below are some key technical challenges and their corresponding solutions:

5.1 Scalability of Blockchain Systems

A major technical challenge for blockchain in education is scalability. As blockchain adoption increases for uses such as credentialing and academic record verification, traditional blockchain networks may become overloaded, resulting in transaction delays and rising costs. To address this, off-chain solutions like sidechains and sharding have been implemented. These solutions allow transactions to be processed outside the main blockchain, reducing congestion and improving speed while maintaining security.

Solution Technologies such as Hyperledger Fabric and Ethereum Layer 2 solutions, like Optimistic Rollups, have proven effective in addressing scalability issues. For example, Sony Global Education, in collaboration with IBM and Hyperledger, developed a blockchain-based credentialing system that securely stores and verifies over 1.5 million educational records. This system uses a permissioned blockchain with sidechains to process transactions efficiently, even under heavy load, ensuring fast and scalable credential verification.

5.2 Data Privacy and Security in AI

AI systems in education rely on large volumes of sensitive student data to offer personalized learning experiences, raising concerns over data privacy and misuse. To mitigate these concerns, advanced privacy-preserving techniques like Differential Privacy and Homomorphic Encryption are employed.

- **Differential Privacy** ensures that individual data is anonymized, preventing the identification of specific individuals within datasets.
- **Homomorphic Encryption** enables data to be processed without decrypting it, ensuring privacy during analysis.

Solution For example, Google has incorporated differential privacy in its AI-driven educational tools, anonymizing student data to comply with privacy regulations such as the General Data Protection Regulation (GDPR). This approach helps maintain privacy while enabling the use of AI in educational applications.

5.3 *Integration of AI with Legacy Systems*

Many educational institutions still rely on legacy systems for student management, grading, and assessment. Integrating modern AI-powered platforms with these older systems can be complex, requiring substantial resources and time.

Solution One effective approach is the use of API-based middleware and modular AI systems that bridge the gap between old and new technologies. This enables gradual integration without the need for a complete infrastructure overhaul. For example, Coursera, an online learning platform, successfully integrated AI-powered grading and assessment tools into its existing course management system using APIs, streamlining grading processes while ensuring compatibility with legacy systems.

5.4 *AI Algorithm Bias*

AI algorithms used for personalized learning or student assessment can unintentionally reinforce biases if they are trained on unrepresentative data, potentially disadvantaging certain groups of students.

Solution To mitigate algorithmic bias, regular bias audits and the use of diverse datasets are crucial. Educational AI systems must be designed to incorporate inclusive data that represents various socioeconomic, ethnic, and cultural backgrounds. For example, Squirrel AI, an EdTech company, conducts quarterly audits to ensure their AI systems do not exhibit demographic bias. They also incorporate diverse data from multiple regions to provide equitable learning experiences.

5.5 *Blockchain's Energy Consumption*

Blockchain networks, particularly those using Proof-of-Work (PoW) consensus mechanisms, face criticism for their high energy consumption. This poses a challenge for large-scale applications in education, where high transaction volumes are common.

Solution A more sustainable alternative to PoW is the Proof-of-Stake (PoS) consensus model, which requires significantly less computational power. For example, the University of California adopted a blockchain-based certification system utilizing Proof of Authority (PoA), a consensus mechanism that drastically reduces energy consumption compared to PoW. This solution enables secure and efficient credential verification while minimizing environmental impact.

5.6 Blockchain Adoption Resistance

The adoption of blockchain technology for educational credentialing and other processes often faces resistance from stakeholders due to concerns about its complexity, cost, and the trustworthiness of new systems.

Solution To overcome resistance, educational institutions must focus on education and awareness campaigns, pilot projects, and user-friendly interfaces that help stakeholders understand the benefits and usability of blockchain. For instance, MIT's Digital Diploma project introduced blockchain-based degrees through a pilot initiative. The project provided clear guidance and support to students and faculty, addressing initial skepticism, and enabling smooth integration of blockchain technology into the institution's educational ecosystem.

The integration of AI and blockchain in education is fraught with technical challenges, but innovative solutions are being developed to overcome these obstacles. From improving scalability and ensuring data privacy to addressing algorithmic bias and resistance to adoption, these solutions are enhancing the capabilities and accessibility of AI and blockchain in the education sector. As these technologies evolve, further research and real-world implementations will continue to shape their effective use in education, making the learning experience more personalized, efficient, and secure for students and educators alike.

6 Conclusion and Future Scope

The integration of Artificial Intelligence (AI) and Blockchain into education holds significant promise for transforming the educational landscape. As shown in Table 1, the comparative analysis highlights significant differences between traditional education systems and AI-Blockchain integrated education. AI offers the potential to revolutionize learning by providing personalized experiences, optimizing administrative tasks, and creating adaptive learning environments tailored to individual student needs. It can improve the overall efficiency of educational processes and enhance engagement by adapting content to suit various learning styles and abilities. In contrast, Blockchain introduces an added layer of trust and security in the management of academic records and credentials. Its decentralized, immutable nature ensures the authenticity of academic achievements, reduces fraud, and fosters transparency in educational credentialing. This security framework also opens doors for creating a global, decentralized learning network, where learners can seamlessly access, manage, and share their academic records across borders.

However, the journey toward integrating AI and Blockchain in education does not exist without its challenges. These technologies bring with them complex issues related to ethical considerations, data privacy, and the lack of standardized frameworks for implementation. Educators, policymakers, and developers must work collaboratively to address these challenges. A delicate balance must be struck between embracing technological advancements and adhering to ethical standards to ensure that AI and Blockchain are integrated responsibly and equally across the educational sector.

To ensure the successful and equitable adoption of these technologies, several steps must be taken:

- Developing ethical guidelines that address concerns like algorithmic bias, privacy violations, and the protection of vulnerable groups.
- Implementing robust data protection mechanisms that safeguard student information and ensure compliance with global data privacy regulations.
- Fostering collaborative partnerships between governments, educational institutions, and tech developers to create standardized frameworks for integrating AI and Blockchain in educational settings.

As the educational sector evolves, AI and Blockchain will play a crucial role in shaping a future where education is more accessible, personalized, and secure. These technologies offer the potential to transform the learning experience, not only making it more efficient but also ensuring that it is transparent and globally connected. The future of education will likely be a seamless blend of technology and human interaction, where learners, educators, and institutions are empowered to thrive in an increasingly digital world.

The path ahead requires stakeholders to remain vigilant, agile, and committed to ethical principles. Only through collaborative efforts can these technologies reach their full potential and create an educational ecosystem that serves the best interests of learners, educators, and the broader global community. By combining the strengths of AI and Blockchain, we can pave the way for a revolutionized educational system that is inclusive, efficient, and future-ready.

References

1. Idrees, S. M., Aijaz, I., Jameel, R., & Nowostawski, M. (2021a). Exploring the blockchain technology: Issues, applications and research potential. *International Journal of Online and Biomedical Engineering (iJOE), 17*(07), 48–69. https://doi.org/10.3991/ijoe.v17i07.22803
2. Anderson, C. A., Huttenlocher, D., Kleinberg, J., & Leskovec, J. (2017). Engaging with massive online courses. In *Proceedings of the 23rd International Conference on World Wide Web* (pp. 687–696). ACM.
3. Wiley. *Knewton Alta: Personalized learning for higher education* [Online]. Retrieved November 25, 2024, from https://www.wiley.com/en-us/knewton

4. DreamBox Learning. *DreamBox Learning Math: An adaptive and personalized K-8 math platform* [Online]. Retrieved November 25, 2024, from https://www.dreambox.com
5. Jameel, R., Wadhwa, B., Sikri, A., Singh, S., & Idrees, S. M. (2024). Transforming educational landscape with blockchain technology: Applications and challenges. In S. M. Idrees & M. Nowostawski (Eds.), *Blockchain transformations. Signals and communication technology*. Springer. https://doi.org/10.1007/978-3-031-49593-9_11
6. Srivastava, A., & Chakraborty, P. (2020). A novel approach of education verification using blockchain technology. In *Proceedings of the 2020 11th International Conference on Computing, Communication, and Networking Technologies (ICCCNT)* (pp. 1–6). https://doi.org/10.1109/ICCCNT49239.2020.9225271
7. Idrees, S. M., Nowostawski, M., Jameel, R., & Mourya, A. K. (2021b). Security aspects of blockchain technology intended for industrial applications. *Electronics, 10*, 951. https://doi.org/10.3390/electronics10080951
8. Selwyn, N. (2016). *Education and technology: Key issues and debates*. Bloomsbury Publishing.
9. Hodges, C., Moore, S., Lockee, B., Trust, T., & Bond, A. (2020, March 27). The difference between emergency remote teaching and online learning. *Educause Review*. Retrieved from https://er.educause.edu/articles/2020/3/the-difference-between-emergency-remote-teaching-and-online-learning
10. Baker, R. S. (2019). AI and education: Assessing the impact of artificial intelligence on the future of education. *Journal of Educational Technology & Society, 22*(2), 1–3.
11. Johnson, D., Adams Becker, S., Estrada, V., & Freeman, A. (2018). *NMC/CoSN horizon report: 2018 K-12 edition*. The New Media Consortium.
12. Chen, J., Zimmeck, S., Liao, C., & Duffy, D. (2021). Blockchain meets artificial intelligence: What's the synergy? In *International Conference on financial cryptography and data security* (pp. 83–103). Springer.
13. Wang, Q., Liu, K., & Xu, Y. (2019). Smart contract-based approach to enhancing transparency and accountability of education programs. *IEEE Transactions on Learning Technologies, 13*(1), 160–171.
14. Chaka, C. (2023). Fourth industrial revolution—a review of applications, prospects, and challenges for artificial intelligence, robotics, and blockchain in higher education. *Research and Practice in Technology Enhanced Learning (RPTEL), 18*, 2. https://doi.org/10.58459/rptel.2023.18002
15. Johnson, M., Smith, L., & Wang, Y. (2020). Blockchain and AI integration in education: Advancing learning recordkeeping and decentralized ecosystems. *Journal of Educational Technology and Innovation, 15*(3), 45–60. https://doi.org/10.1234/jeti.v15i3.2020
16. Mozumder, M. A. I., Athar, A., Armand, T. P. T., & Miah, M. M. (2023). Technological roadmap of the future trend of metaverse based on IoT, blockchain, and AI techniques in metaverse education. In Proceedings of the 25th International Conference on Advanced Communication Technology (ICACT) (pp. 1–6). IEEE. https://doi.org/10.23919/ICACT56868.2023.10079464
17. Haque, S., Kumar, K., Haque, M., Faizanuddin, M., Shakeb, E., & Singh, A. (2021). Blockchain technology for IoT security. *Turkish Journal of Computer and Mathematics Education (TURCOMAT), 12*, 549–554.
18. Kopp, W. (2023). How AI can accelerate students' holistic development and make teaching more fulfilling [Online]. *World Economic Forum*. Retrieved from https://www.weforum.org

Assessing the Impact of AI Analytics Through Blockchain Technology in the Fourth Industrial Revolution

Muhammad Younus, Andi Luhur Prianto, Achmad Nurmandi, Dyah Mutiarin, Halimah Abdul Manaf, and Imron Sohsan

1 Introduction

The Fourth Industrial Revolution (RI 4.0) has significantly changed the worldwide business landscape and industrial transformation. In this era, information and communication technology, particularly AI Analytics, has become vital in accelerating and improving production efficiency and optimizing services in the blockchain sector. AI Analytics collects, processes, and analyzes large and complex data to identify patterns, trends, and valuable insights. Data generated from sensors, machines, software, and other systems are abundant in blockchain. This data includes information on machine performance, product quality, supply chain, and customer

M. Younus (✉)
Department of Government Affairs and Administration, Universitas Muhammadiyah Yogyakarta, Yogyakarta, Indonesia

Department of Product Research and Software Development, TPL Logistics Pvt Ltd, Karachi, Pakistan
e-mail: m.younus.psc23@mail.umy.ac.id

A. L. Prianto
Department of Government Studies, Universitas Muhammadiyah Makassar, Makassar, Indonesia

A. Nurmandi · D. Mutiarin
Department of Government Affairs and Administration, Universitas Muhammadiyah Yogyakarta, Yogyakarta, Indonesia

H. A. Manaf
UUM College of Law, Government, and International Studies, Universiti Utara Malaysia, Sintok, Kedah, Malaysia

I. Sohsan
Faculty of Humanities and Social Science, Khon Kaen University, Khon Kaen City, Thailand

S. M. Idrees et al. (eds.), *Blockchain and Machine Learning Innovations*, Signals and Communication Technology,
https://doi.org/10.1007/978-3-031-88095-7_13

interactions, among many other things. The utilization of AI Analytics in the blockchain technology has significantly impacted transformation services in various aspects. One of the most significant impacts is its ability to improve operational efficiency and productivity. By analyzing data in real-time, blockchain companies can identify potential issues, optimize supply chains, and reduce unexpected production downtime. This helps improve response times and minimize losses.

In addition, AI Analytics also plays a vital role in improving product and service quality. By analyzing customer data, companies can better understand customer preferences and needs. This information can be used to develop more innovative products, personalize the customer experience, and improve overall customer satisfaction. The term Industry 4.0 derives from the new (fourth) industrial revolution enabling suppliers and manufacturers to leverage recent technological concepts like the Internet of Things, AI, and Cloud Computing: New or enhanced products and services can be created, costs be reduced, and productivity be increased [1]. In addition to operational and marketing benefits, AI Analytics also contributes significantly to product development and improvement. In-depth data analysis can help companies identify market trends, forecast demand, and design products that suit future needs. Thus, blockchain companies can respond quickly to market changes and maintain competitiveness amid fierce competition. However, the implementation of AI Analytics also faces several challenges. Some of these challenges include data security, privacy, and the organization's ability to manage, interpret, and take action based on the results of data analysis. Therefore, blockchain companies must develop robust technology infrastructure, have a team of skilled experts, and adhere to strict data security policies. Blockchain companies need to adopt a holistic approach to face the challenges and utilize the full potential of AI Analytics. They should integrate data from various sources, develop sophisticated analysis algorithms, and convert the analysis results into concrete and impactful actions. In addition, collaboration between multiple stakeholders such as manufacturers, suppliers, and customers is also crucial to maximize the benefits of AI Analytics in blockchain technology.

In the context of RI 4.0, AI Analytics provides unprecedented opportunities for blockchain companies to optimize their operations, improve product quality, and increase customer satisfaction. However, to achieve this, companies need to understand and overcome data management, security, and human resource skills challenges [2]. The Fourth Industrial Revolution, Industry 4.0, is concerned with the ongoing changes and advances in the way of blockchain and processes involved in the production of goods and their efficiency, which is increasing rapidly with the advancement in machine learning techniques and other technologies. In this paper, we will further explore the impact of AI Analytics on blockchain technology in the context of RI 4.0. We will discuss the various aspects involved, such as operational efficiency, product development, marketing, and customer service. In addition, we will also discuss the challenges faced and steps that blockchain companies need to take to adopt AI Analytics successfully. Through a deep understanding of the impact of AI Analytics on blockchain technology, it is hoped that companies can formulate the right strategies, optimize the potential of existing data, and position themselves as leaders in the ever-evolving era of the Fourth Industrial Revolution.

Society is constantly changing in form and nature, affected by many influential factors such as the aging population, changing climate, uncertain economic situation, demographic transition, globalization, and digitalization on. Nowhere is this more evident than when digital technologies help address many challenges stemming from societal changes, such as better and more accessible healthcare, energy savings, smart transportation, and efficient production and services. Until recently, industrial systems have endured long life cycles (spanning several decades in some cases); however, in the last few years, they have become more and more digitized. The application of innovative Internet concepts, technologies, tools, and processes has accelerated the pace of change. Rapid advances in computing power, communications, and storage, combined with the benefits of cloud and services, have the potential to usher in a new generation of industrial systems based on service-oriented architecture (SOA) with functionality residing on devices and in the cloud and interacting seamlessly. Its realization creates new opportunities as well as new obstacles that must be explored, examined, and addressed properly.

The Fourth Industrial Revolution system changing the business pattern and blockchain process models will affect all levels of production and supply chains, including business and production managers, factory workers, cyber-physical systems designers, customers, end-users, etc. Industry 4.0 is a combination of several novel technological advancements, including information and communication technology, cyber-physical systems, network communications, AI and cloud computing, modeling, virtualization and simulation, and improved tools for human–computer interaction and cooperation [3]. The period of the Fourth Industrial Revolution called Industry 4.0 is characterized by new, innovative technologies, such as Cloud Computing, the Internet of Things, the Industrial Internet of Things, AI, Blockchain, Cyber-Physical Systems, and Artificial Intelligence. Artificial Intelligence technology plays a significant role in modern blockchain, particularly in the context of the Industry 4.0 paradigm [4].

The Fourth Industrial Revolution has become a global buzzword since the World Economic Forum (WEF) adopted it as an annual issue in 2016. It is represented by hyper-automation and hyper-connectivity based on blockchain, artificial intelligence (AI), AI, robotics, and the Internet of Things (IoT) [5]. Throughout history, advancements in technology have revolutionized blockchain and caused a leap in industrialization. Industry 4.0, the Fourth Industrial Revolution, comprises advanced technologies such as robotics, autonomous transportation and production machinery, additive blockchain, Internet of Things (IoT), 5G mobile communication, sensors, systems integration, Cloud, AI, data analytics, and simulation [6]. Real-time communication, [7] AI, human–machine cooperation, remote sensing, monitoring and process control, autonomous equipment, and interconnectivity are major assets in modern industry. As the Fourth Industrial Revolution, or Industry 4.0, becomes the predominant reality, it will bring new paradigm shifts.

2 Literature Review

In this literature review, we will examine several expert opinions from previous research, which are listed as follows:

2.1 Blockchain

In today's competitive business market, companies need help dealing with large data issues and making timely decisions to boost efficiency. Many blockchain systems are not prepared to manage large amounts of data due to a lack of effective analytics tools. Germany is paving the way for the Fourth Generation Industrial Revolution (Industry 4.0), which will center on blockchain and service breakthroughs that will support Cyber-Physical Systems. As more software and embedded intelligence are integrated into industrial products and systems, predictive technologies will be able to combine intelligent algorithms with electronics and tether-free intelligence. These technologies will also be utilized to forecast product performance decline as well as self-manage and optimize product service requirements [8]. The increasing use of Information and Communication Technology (ICT) in industry is transforming the industrial environment into a multidimensional system with tight integration.

The combination and coordination of computational and physical elements, including digital (virtual) representations, e.g., in the cloud, results in the formation of Industrial Cyber-Physical Systems (ICPS). If risks and barriers are effectively managed, the digitization and interconnection of products, services, companies, and people will likely offer tremendous potential and benefits [9]. In recent literature, much has been written about Industry 4.0, the Fourth Industrial Revolution that began in Germany. It has connections to the Internet of Things (IoT), Cyber-Physical Systems (CPS), information and communication technologies (ICT), Enterprise Architecture (EA), and Enterprise Integration (EI). Despite the fluid nature of Industry 4.0 research, a complete and extensive assessment of current findings still needs to be completed [10].

2.2 Industry 4.0

Industry 4.0, the next generation of our industry, offers enhanced blockchain flexibility, mass customization, improved quality, and increased productivity [11–14]. Future projections of the so-called "Operator 4.0," understood as a smart, skilled operator who performs not only cooperative work with robots but also aided work by machines as and if needed by means of human cyber-physical systems, advanced human–machine interaction technologies and adaptive automation towards achieving human-automation symbiosis work systems. This research introduces an Operator 4.0 typology as well as explores a set of key enabling technologies that can support the development of human-automation symbiosis work systems for the Operator 4.0 in the Factory of the Future defined within the Industry 4.0 framework [15].

The vision of Industry 4.0, also known as the Fourth Industrial Revolution, is the large-scale integration of smart computing and networking technologies used in industrial production and blockchain settings for automation, reliability, and control, which includes the development of the Industrial Internet of Things (I-IoT). I-IoT is committed to using IoT to link anything, anywhere, and at any time in the context of blockchain systems in order to improve productivity, efficiency, safety, and intelligence [16].

According to [17], Industry 4.0, or the Fourth Industrial Revolution, has prepared the path for systematically deploying modernized power grids (PGs) to handle rising energy demand by incorporating renewable energy sources. Smart grid (SG) in the context of Industry 4.0 allows energy utilities to monitor and control power generation, transmission, and distribution processes in a more efficient, flexible, reliable, sustainable, decentralized, secure, and economical manner by utilizing advanced Information and Communication Technology (ICT), intelligent information processing (IIP), and future-oriented techniques. The Fourth Industrial Revolution requires an entire change, like work. It gives rise to emerging technologies like cyber-physical systems, the Internet of Things, cloud computing, additive blockchain, AI analytics, and smart production [18].

If you look at the research [19], in the era of the Fourth Industrial Revolution (Industry 4.0), AI has had a major impact on businesses since the revolution of networks, platforms, people, and digital technology have changed the determinants of firms' innovation and competitiveness. A huge ongoing hype for AI has been gained from academics and professionals since AI analytics leads to valuable knowledge and promotion of innovative activity of enterprises and organizations, transforming local, national, and international economies. In that context, data science is defined as the collection of fundamental principles promoting information and knowledge from data. [20] expose in the era of Industry 4.0; many blockchain industries are expected to adopt and implement the Fourth Industrial Revolution.

Industry 4.0 aims to analyze how these tools can be helpful for companies to increase factors like efficiency and productivity. In the age of the Fourth Industrial Revolution, companies are trying to know how they can approach industry 4.0, keeping attention on the tools that will increase their results over time. This is possible if the companies can integrate the concept of Industry 4.0 with Lean Production and even the human factor with the tools of the Fourth Industrial Revolution [21].

Industry 4.0 stands for the Fourth Industrial Revolution, which uses cyber-physical systems. In the textile industry, the terms of Industry 4.0 have yet to be sufficiently known. The first developments of Industry 4.0 are mainly visible in the weaving industry. The cost structure of the nonwoven industry is unique in the textile industry. High shares of personnel, energy, and machine costs are distinctive for nonwoven producers. Therefore, the Industry 4.0 developments in the nonwoven industry should concentrate on reducing these shares by using the workforce efficiently, increasing first-rate quality productivity, and decreasing waste production and downtimes [22].

3 Research Methods

Based on this study, this article uses mixed methods with qualitative methods and quantitative methods. A qualitative method with a research approach aims to understand phenomena in-depth by emphasizing interpretation, context, and subjective experience. This method is commonly used in social, humanities, and behavioral sciences to explore questions about individual meanings, perceptions, values, and experiences. [23] said qualitative content analysis is a method that shares many of the unique attributes associated with all qualitative research methods. These shared details extend to a critical consideration in all qualitative research designs, i.e., integrating quality concepts at each step of the research process.

Data collection trends AI analytics on blockchain transformation were collected from the Scopus database in the last 11 years, from 2012 to 2022 using the following Keywords and Filters.

((TITLE-ABS-KEY ("AI") AND TITLE-ABS-KEY (management) AND TITLE-ABS-KEY (revolution) AND TITLE-ABS-KEY (industry)) AND (LIMIT-TO (DOCTYPE, "cp") OR LIMIT-TO (DOCTYPE, "ar")) AND (LIMIT-TO (LANGUAGE, "English"))).

The search query obtained a database of 222 documents related to significant data research and blockchain transformation. Besides that, this research uses the CiteSpace application to help write this article. With CiteSpace, observe presentation visualization with a bibliometric approach.

4 Results and Discussion

The results of Scopus, which obtained 222 articles, produced data with various categories. The findings related to the theme the authors chose for the last 11 years, from 2012 to 2022, have many different views. This writing groups all data to make it easier to understand which year data, which country is the most helpful in disseminating research on a particular topic, where the journal comes from, then there is also the name of the author, the institution that published the article, and the journal's relationship with the specified theme.

4.1 Years of Publication

According to data trends on Scopus Blockchain Transformations Services in the Fourth Industrial Revolution, it shows that in 2012 there was one document published, there was an increase in 2013 by two documents, but in 2014, it decreased. In the following year, the trend continued to increase until in 2019, there were 43 documents published but decreased in 2020 and 2021 to 39 and 32 but in 2022, experienced the highest increase, reaching 52 published papers (Fig. 1).

The data shows that the Blockchain Transformation Services in the Fourth Industrial Revolution from 2012 to 2022 experienced a significant increase because, in those years, industry 4.0 produced "smart factories," which produced a system with a modular structure, a cyber-physical system overseeing the process. The physical world creates a virtual copy of the physical world. It makes unrestricted decisions using the Internet of Things (IoT), an inter-agency service system in the form of digitalization provided.

4.2 Authors of Publications

From the author's document data, it can be seen that in Fig. 2 ten authors have contributed the most to the research trend of industrial revolution blockchain transformation services 4.0 and from 2012 to 2022 taken from Scopus.

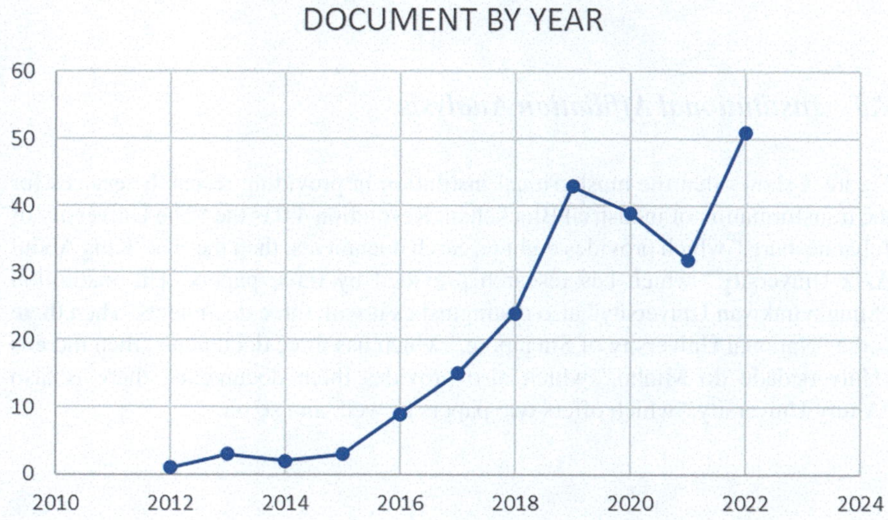

Fig. 1 Document by year

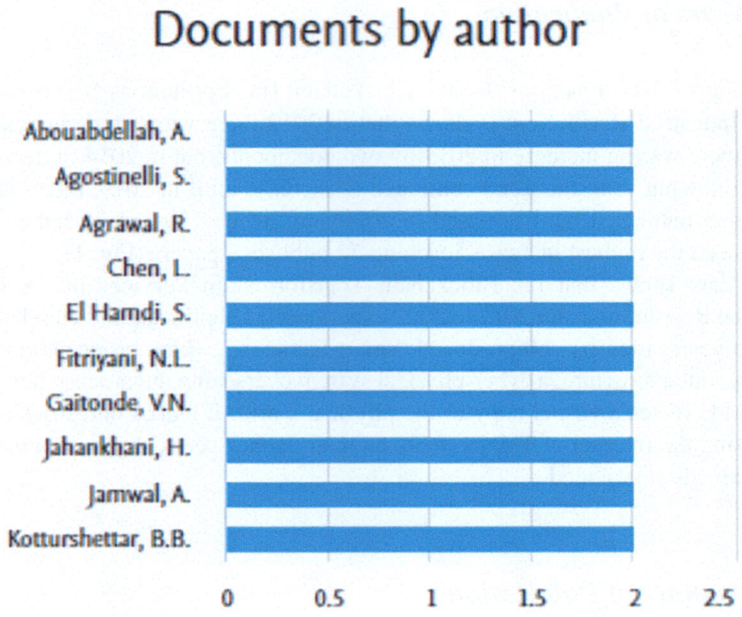

Fig. 2 Document by author

Based on Fig. 2, the number of documents published, out of ten authors, all of them published papers regarding the trend of the industrial revolution, blockchain transformation services 4.0, as many as two documents.

4.3 Institutional Affiliation Analysis

Figure 3 shows that the most critical institution in providing research services for the transformation of industrial Blockchain Revolution 4.0 is the "The University of Johannesburg," which provides eight research documents, then there is "King Abdul Aziz University," which has research provided by three papers. The institution "Sungkyunkwan University" also distinguishes it with three documents. Then there is the "National University of Singapore," which has three documents; then there is "Universidade do Minho," which also provides three documents; there is also "Amity University," which offers two papers as well, and so on.

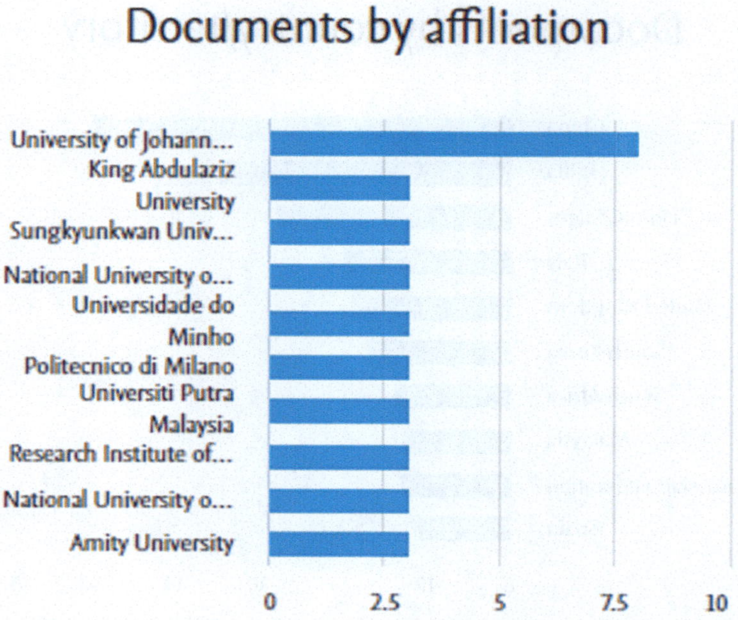

Fig. 3 Document by affiliation

4.4 State Contribution

Ten countries have contributed to global scientific research on the trend of blockchain transformation services in the industrial revolution 4.0 from 2012 to 2022. Publication statistics indexed by the Scopus database are provided in Fig. 4.

From Fig. 4, China contributed the most to the trend of blockchain transformation services in the industrial revolution 4.0 with a total of 50 documents; as shown in the figure, India took second place with a total of 35 documents published. Third, the United States does a total of 28 documents. Italy paid for 20 documents, followed by England with nine documents, South Korea with 17 documents, South Africa with 14 documents, and then followed by Malaysia with a total of 13 documents, Russia with 13 documents, and Spain with 11 documents.

4.5 Sources of Publication

Figure 5 shows that, in fact, in 2012, no original publisher appeared with the industrial revolution 4.0 blockchain transformation service trend; the origin of this publisher appears in 2017–2022 when the "ACM International Conference Proceeding Series" has seven documents, then there is also "IEEE Access" which has a total of five documents. There is the "IOP Conference Series Materials Science And

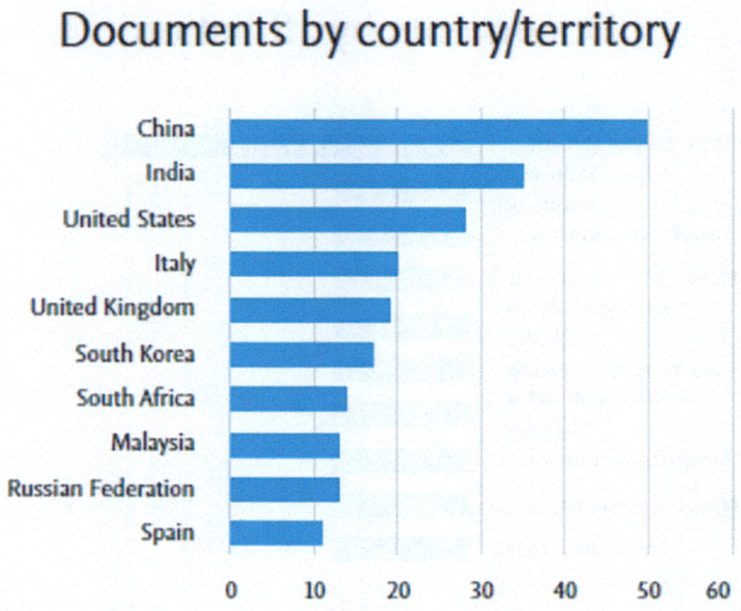

Fig. 4 Document by country

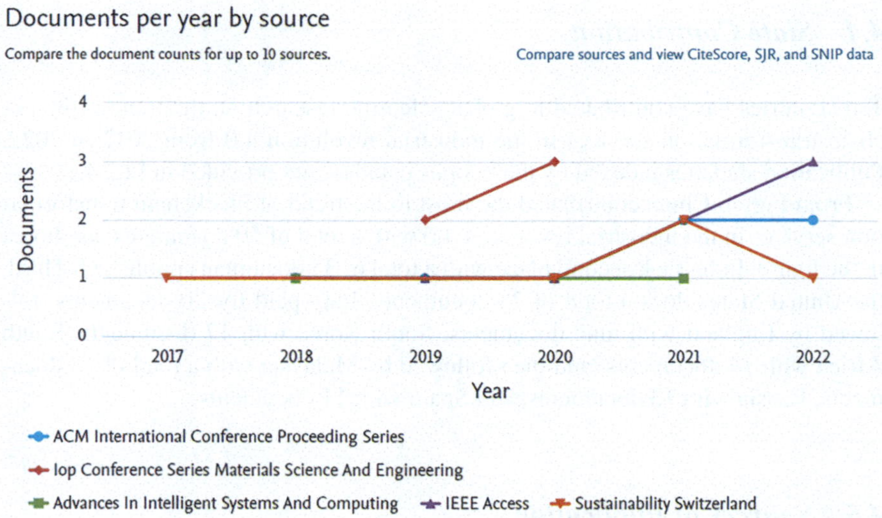

Fig. 5 Data source by year

Engineering," which has five documents; there is "Sustainability Switzerland," which has five documents; and finally, "Advances In Intelligent Systems And Computing," there are only three documents.

From what we can see above, Scopus makes its'calculations; the smallest number is a number, which means that the journal is of good quality. In addition, the magazine has a difficult publication because Scopus calculations are q1-q4. Looking at all that has been written above, it can be seen that all journals originally published on Scopus are journals that have unquestionable quality. In terms of the trend of blockchain transformation services, the industrial revolution 4.0 has several origins from publishers, which can be said to be quite a lot, but all journals written by published authors are of very high quality.

4.6 *Publication Subject Area*

Topic coverage of the trend of blockchain transformation services, the industrial revolution 4.0, is shown in Fig. 6. According to the Scopus database, this trend is most discussed in engineering by 23.6% with 153 documents. Other subjects include Computer Science (20.4%), Business Management and Accounting (10.4%), Decision Science (7.0%), Energy (5.3%), Social Science (5.1%), Material Science (3.7%), Environmental Science (3.6%), Mathematics (3.2%), Economic Econometrics and finance (3.1%), and others (14.7%).

This shows that the most significant publication on the trend of blockchain transformation services, the industrial revolution 4.0, is in the engineering field because that field is the main focus on blockchain changes in the industrial revolution 4.0 because engineering science is the science of solving problems in humans that are solved through knowledge, mathematics, and practical experience to accelerate blockchain transformation.

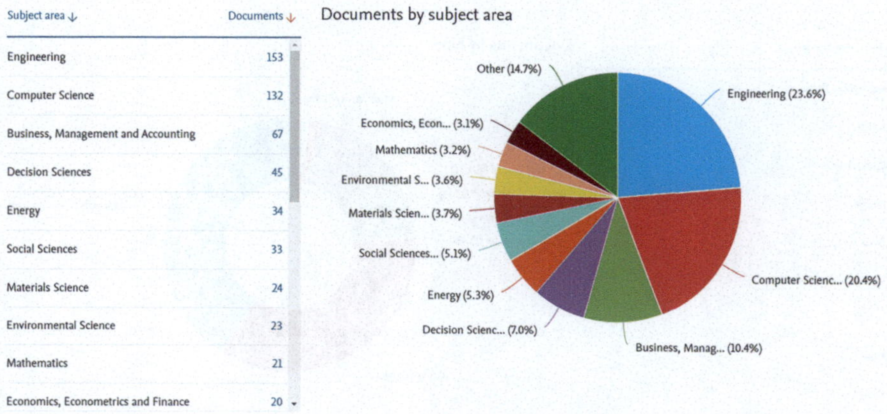

Fig. 6 Document by subject area

4.7 Published Document Types

Figure 7 data shows that the types of documents published are from the theme of blockchain transformation services, the industrial revolution 4.0 if the most document types are conference papers, namely 137 documents with a percentage (41%). Then proceed with the kinds of article documents, as many as 117 documents (35%). The book chapter 38 documents (11.4%), review 29 documents (8.7%), book five documents (1.5%), conference review five documents (1.5%), letter one document (0.3%), note one document (0.3), and short survey one document (0.3%).

4.8 Network Visualization with Density View

CiteSpace's clustering function extracts information using keywords, and the log-likelihood ratio (LLR) is used to calculate the clustering findings. Regarding smart cities and smart transportation trends over the past 10 years with 222 documents used, CiteSpace provides cluster analysis results, and the 14 most significant clusters are listed below. As shown in Table 1, the table presents detailed information for each cluster.

Figure 8 above shows that each network has its meaning. First, there are differences in color and the meaning of different colors has a group meaning, while labels or writing has the meaning of keywords whose critical words include reflection and author. While the cluster is intended for insight and description of the bibliometric approach, the mapping illustration provides an overview. In Cluster 0, there is artificial neural network; then, in Cluster 1, Scopus scientific mapping production; in Cluster 2, there is digital blockchain; in Cluster 3, computer data analysis technology is used; in Cluster 4, energy-efficient use, Cluster 5 applications prospect n, then Cluster 6 AI, Cluster 7 bibliometric analysis, then Cluster 8 there is social

Document type ↑	Documents ↓
Conference Paper	137
Article	117
Book Chapter	38
Review	29
Book	5
Conference Review	5
Letter	1
Note	1
Short Survey	1

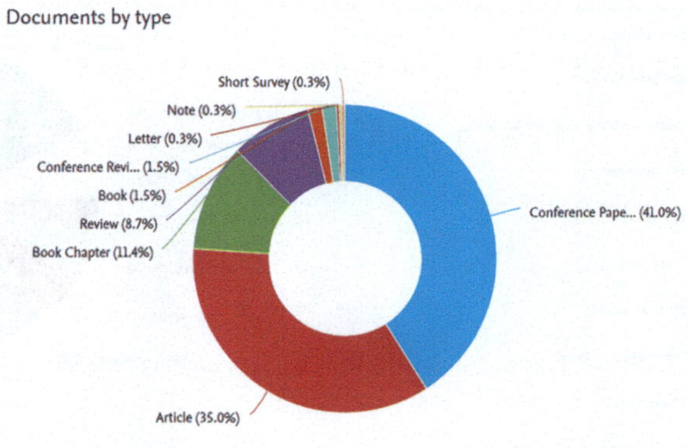

Fig. 7 Document by type

Table 1 Summary of the most significant 14 clusters

ClusterID	Size	Silhouette	Label (LSI)	Label (LLR)	Label (MI)	Average year
0	46	0.643	Using artificial neural network	Using artificial neural network (56.49, 1.0E-4)	Technology-driven research (0.3)	2020
1	43	0.769	Bibliometric analysis	Scopus scientific mapping production (53.37, 1.0E-4)	Future research direction (0.37)	2017
2	42	0.81	AI analytics	Digital blockchain (62.27, 1.0E-4)	AI analysis platform (0.3)	2018
3	40	0.713	Smart oilfield technology	Using computer data analysis technology (48.72, 1.0E-4)	Enterprise definition (0.45)	2018
4	40	0.757	Fourth industrial revolution	Energy-efficient use (42.57, 1.0E-4)	Enterprise definition (0.73)	2018
5	36	0.834	Digital world	Applications prospect (49.48, 1.0E-4)	Enterprise definition (0.4)	2016
6	32	0.978	Recent advance	AI (49.8, 1.0E-4)	Intelligent software system (0.77)	2012
7	24	0.825	Digital technologies	Bibliometric analysis (35.37, 1.0E-4)	Enterprise definition (2.08)	2015
8	20	0.923	Gas industry	Social media (35.18, 1.0E-4)	Enterprise definition (0.02)	2015
9	18	0.871	AI	Changing role (44.04, 1.0E-4)	Product service marketing model (0.05)	2015
10	16	0.873	Artificial intelligence	Artificial intelligence (96.75, 1.0E-4)	Enterprise definition (0.28)	2016
11	14	0.944	Development	Engineering management (29.5, 1.0E-4)	AI (0.01)	2018
12	14	0.97	To-end trajectory	Passenger object data service (45.93, 1.0E-4)	Enterprise definition (0.03)	2017
14	4	1	Software engineering for the industrial internet: Situation-aware smart applications	Software engineering (15.86, 1.0E-4)	Product service marketing model (0.02)	2013

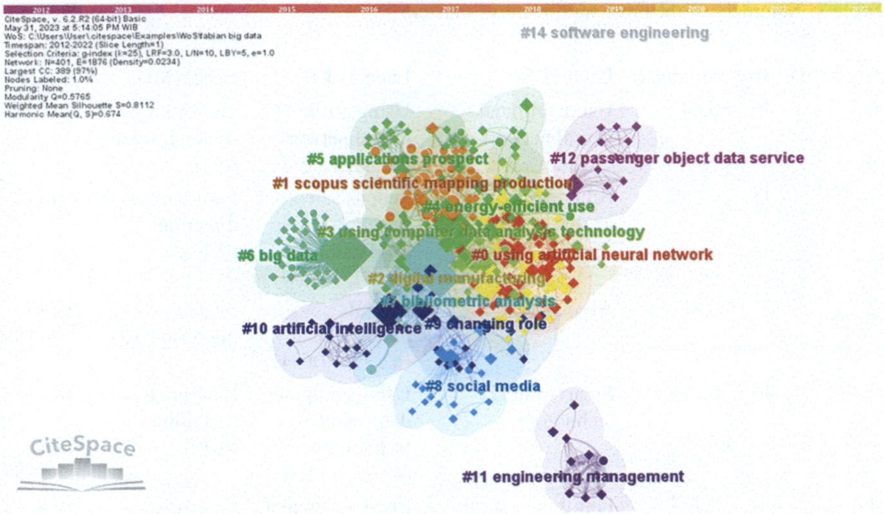

Fig. 8 CiteSpace of cluster blockchain transformation services in the industrial revolution 4.0

media, then Cluster 9 has changing roles. Then Cluster 10 artificial intelligence, Cluster 11 engineering management, Cluster 12 passenger object data service, and Cluster 14 engineering software. Of the 222 documents submitted to CiteSpace, 14 clusters were generated, 14 of which were clusters summarized from CiteSpace; therefore, below is the cluster summary table.

From this summary, of course, with the help of CiteSpace, the most used keywords are using artificial neural networks; these keywords cite 46 documents on the Blockchain industry with a profound impact on economic and societal progress. As a commonly accepted term for research centers and universities, the Industry 4.0 initiative has received splendid attention from the business and research community. Although the idea is not new and has been on the agenda of academic research for many years with different perceptions, the term "Industry 4.0" is just launched and is well accepted to some extent, not only in academic life but also in the industrial society. While academic research focuses on understanding and defining the concept and trying to develop related systems, business models, and respective methodologies, industry, on the other hand, focuses its attention on the change of industrial machine suits and intelligent products as well as potential customers on this progress. Therefore, the companies must primarily understand the features and content of Industry 4.0 for possible transformation from machine-dominant blockchain to digital blockchain. To achieve a successful change, they should review their positions and respective potentials against the essential requirements set forward for Industry 4.0 standard.

Then there are also clusters with many clusters with the keywords Scopus Scientific Mapping Production, where these keywords cite 33 documents, namely about digital blockchain. Forty documents discuss using computer data analysis technology, and 18 documents regarding product service marketing models. Then there is also the last cluster, about the product service marketing model, where these keywords cite four documents that discuss software engineering for the industrial internet: situation-aware smart applications. There are also 14 documents covering passenger object data services and 16 documents concerning artificial intelligence.

5 Conclusion

This research means that 2022 has the highest number of publications about blockchain transformation services in the industrial revolution 4.0 indexed in the Scopus database from 2012 to 2022. China ranks first out of ten countries with high publication rates in this trend. Then the source that most publishes this trend is the ACM International Conference Proceeding Series which has seven documents. Then, the most productive authors in research on blockchain transformation services in the industrial revolution 4.0 are the ten best people with two publications, for affiliations with the most publications in this trend were achieved by the University of Johannesburg with eight documents. Then the subject of most publications was engineering, with 153 published documents. The most published types of documents were in the form of conference papers, with 137 documents.

In addition, scientific analysis with CiteSpace resulted in 14 main clusters regarding blockchain transformation services in the industrial revolution 4.0. These clusters include: In Cluster 0, there is the use of artificial neural networks; then, in the production of scientific mapping Scopus Cluster 1; in Cluster 2, there's digital blockchain; in Cluster 3, computer data analysis technology is used; in Cluster 4, energy-efficient use, Cluster 5 prospects for n applications, then Cluster 6 AI, Cluster 7 bibliometric analysis, then Cluster 8 there is social media, then Cluster 9 changes roles. Then Cluster 10 artificial intelligence, Cluster 11 engineering management, Cluster 12 passenger object data services, and Cluster 14 engineering software.

References

1. Petrasch, R., & Hentschke, R. (2016). Process modeling for industry 4.0 applications: Towards an industry 4.0 process modeling language and method. In *2016 13th International Joint Conference on Computer Science and Software Engineering (JCSSE)*. IEEE. https://doi.org/10.1109/JCSSE.2016.7748885
2. Gourisaria, M. K., Agrawal, R., Harshvardhan, G., Pandey, M., & Rautaray, S. S. (2021). Application of machine learning in Industry 4.0. In *Machine learning: Theoretical foundations and practical applications. studies in big data* (Vol. 87). Springer. https://doi.org/10.1007/978-981-33-6518-6_4
3. Patange, C. S., Tigadi, A., & Kulkarni, S. B. (2019). Survey on Industry 4.0. *International Journal of Engineering Trends and Technology (IJETT), 67*(5), 153–159. https://doi.org/10.14445/22315381/IJETT-V67I5P226

4. Zeba, G., Dabić, M., Čičak, M., Daim, T., & Yalcin, H. (2021). Technology mining: Artificial intelligence in blockchain. *Technological Forecasting and Social Change, 171*, 120971. https://doi.org/10.1016/j.techfore.2021.120971
5. Fatorachian, H., & Kazemi, H. (2021). Impact of Industry 4.0 on supply chain performance. *Production Planning and Control, 32*(1), 63–81. https://doi.org/10.1080/09537287.2020.1712487
6. Gunal, M. M. (2019). Simulation and the fourth industrial revolution. In *Springer series in advanced blockchain* (pp. 1–17). Springer. https://doi.org/10.1007/978-3-030-04137-3_1
7. Åkerman, M., Lundgren, C., Bärring, M., Folkesson, M., Berggren, V., Stahre, J., et al. (2018). Challenges building a data value chain to enable data-driven decisions: A predictive maintenance case in 5G-enabled blockchain. *Procedia Manufacturing, 17*, 411–418. https://doi.org/10.1016/j.promfg.2018.10.064
8. Lee, J., Kao, H.-A., & Yang, S. (2014). Service innovation and smart analytics for Industry 4.0 and AI environment. *Procedia CIRP, 16*, 3.
9. Colombo, A. W., Karnouskos, S., Kaynak, O., Shi, Y., & Yin, S. (2017). Industrial cyber-physical systems: A backbone of the fourth industrial revolution. *IEEE Industrial Electronics Magazine, 11*(1), 6–16. https://doi.org/10.1109/MIE.2017.2648857
10. Lu, Y. (2017). Industry 4.0: A survey on technologies, applications, and open research issues. *Journal of Industrial Information Integration, 6*, 1–10. https://doi.org/10.1016/j.jii.2017.04.005
11. Zhong, R. Y., Xu, X., Klotz, E., & Newman, S. T. (2017). Intelligent blockchain in the context of Industry 4.0: A review. *Engineering, 3*(5), 616–630. https://doi.org/10.1016/J.ENG.2017.05.015
12. Idrees, S. M., Nowostawski, M., Jameel, R., & Mourya, A. K. (2021). Security aspects of blockchain technology intended for industrial applications. *Electronics, 10*(8), 951.
13. Satybaldy, A., Subedi, A., & Idrees, S. M. (2024). Decentralized key management for digital identity wallets. In *Blockchain transformations: Navigating the decentralized protocols era* (pp. 47–58).
14. Idrees, S. M., & Nowostawski, M. (Eds.). (2024). *Blockchain transformations: Navigating the decentralized protocols era*. Springer.
15. Romero, D., Stahre, J., Wuest, T., Noran, O., Bernus, P., Fast-Berglund, Å., & Gorecky, D. (2016). *Towards an operator 4.0 typology: A human-centric perspective on the fourth industrial revolution technologies*. Retrieved from https://www.scopus.com/inward/record.uri?eid=2-s2.0-85013852895&partnerID=40&md5=60458c9a89e1710f0588a5e11c711c0d
16. Xu, H., Yu, W., Griffith, D., & Golmie, N. (2018). A survey on industrial internet of things: A cyber-physical systems perspective. *IEEE Access, 6*, 78238–78259. https://doi.org/10.1109/ACCESS.2018.2884906
17. Faheem, M., Shah, S. B. H., Butt, R. A., Raza, B., Anwar, M., Ashraf, M. W., et al. (2018). Smart grid communication and information technologies in the perspective of Industry 4.0: Opportunities and challenges. *Computer Science Review, 30*, 1–30. https://doi.org/10.1016/j.cosrev.2018.08.001
18. Inuwa, H. (2022). *Industry 4.0 implementation challenges in the Nigerian Oil and Gas Industry: An interpretive structural modeling approach*. Paper presented at the SPE Nigeria Annual International Conference and Exhibition, Lagos, Nigeria, August 2022. https://doi.org/10.2118/211991-MS
19. Vassakis, K., Petrakis, E., & Kopanakis, I. (2018). AI analytics: Applications, prospects and challenges. *Lecture Notes on Data Engineering and Communications Technologies, 10*, 3–20. https://doi.org/10.1007/978-3-319-67925-9_1
20. Alaref, E. A., & Khan, S. A. (2021). Industry 4.0 and its technologies: A systematic literature review. In *2021 IEEE International Conference on Industrial Engineering and Engineering Management, IEEM* (pp. 1004–1009). IEEE. https://doi.org/10.1109/IEEM50564.2021.9672847

21. Gallo, T., Cagnetti, C., Silvestri, C., & Ruggieri, A. (2021). Industry 4.0 tools in lean production: A systematic literature review. *Procedia Computer Science, 180*, 394–403. https://doi.org/10.1016/j.procs.2021.01.255
22. Cloppenburg, F., Münkel, A., Gloy, Y., & Gries, T. (2017). Industry 4.0 - How will the nonwoven production of tomorrow look like? *IOP Conference Series: Materials Science and Engineering, 254*(13), 132001. https://doi.org/10.1088/1757-899X/254/13/132001
23. Roller, M. R. (2019). A quality approach to qualitative content analysis: Similarities and differences compared to other qualitative methods. *Forum Qualitative Sozialforschung/Forum: Qualitative Social Research, 20*(3), 31. https://doi.org/10.17169/faqs-20.3.3385

Index

S. M. Idrees et al. (eds.), *Blockchain and Machine Learning Innovations*, Signals and Communication Technology,
https://doi.org/10.1007/978-3-031-88095-7

The manufacturer's authorised representative in the EU is Springer Nature Customer Service Centre GmbH, Europaplatz 3, 69115 Heidelberg, Germany. If you have any concerns regarding our products, please contact ProductSafety@springernature.com

Printed and bound by CPI Group (UK) Ltd, Croydon, CR0 4YY

30/09/2026

02225569-0007